Indian Pandanaceae - an overview

Altafhusain Nadaf • Rahul Zanan

Indian Pandanaceae - an overview

Springer

Altafhusain Nadaf
Department of Botany
University of Pune
Pune, India

Rahul Zanan
Department of Botany
University of Pune
Pune, India

ISBN 978-81-322-0752-8 ISBN 978-81-322-0753-5 (eBook)
DOI 10.1007/978-81-322-0753-5
Springer India Heidelberg New York Dordrecht London

Library of Congress Control Number: 2013931657

Printed on acid-free paper

Springer is part of Springer Science+Business Media (www.springer.com)

Authors dedicate this book to their respective parents

Foreword

India can justly lay claim to be the cradle of Pandanaceae taxonomy. The four species of *Pandanus* carefully described and illustrated in van Rheede's *Hortus Malabaricus* in 1679 are the first to be mentioned in the botanical literature. A century later, in 1786, William Roxburg, the Director of the Calcutta Botanical Garden brought van Rheede's Indian screwpine names into line with Carl Linnaeus' new binomial classification system, that was gradually being adopted by botanists all over the world. During the course of the nineteenth and twentieth centuries, several botanists notably Otto Warburg, Ugolino Martelli, Harold St. John and Benjamin Stone worked on the taxonomy of the Indian Screwpines- introducing new synonymies and describing new species, but none of them took on the task to revise the Indian screwpines as a whole. This was mostly due to the difficultly of obtaining good preserved material of this mysterious and fascinating family.

During the past decade, Dr. Altafhusain Nadaf and Dr. Rahul Zanan have undertaken a very demanding challenge: to understand the Indian Pandanaceae as a whole and to advance its taxonomy in a global context, making full use of modern techniques including molecular systematics. The results of this huge and much-needed work are presented in this book: "*Indian Pandanaceae – an overview*".

With this book, the authors present the first complete overview of Indian screwpines more than 300 years after the first *Pandanus* species were recorded by van Rheede from the Malabar region. Dr. Altafhusain Nadaf and Dr. Rahul Zanan have visited the whole sub-continent to collect, study and understand the species in the field. In this book they present the Indian screwpines in unparalleled clarity and detail, dealing with their history, economic importance, molecular phylogenetic and conservation, as well as their taxonomy. I am convinced that this book will serve people with a broad spectrum of interests and it will stimulate future research and interest in this economically important family in India. We could not have hoped for such a thorough and complete overview of the Indian screwpines!

Martin W. Callmander

Preface

While studying the phylogeny of Indian screw pines for the past seven years, we realised that the taxonomy of Indian Pandanaceae is in very unsatisfactory state. Since van Rhedee's time (1679), a number of species have been described that have never been critically compared. The treatment by the taxonomists was very sketchy, as several species were founded on the poorest of specimens and unfortunately some have also been incorrectly or inexactly described leading in the creation of several synonyms and number of species. This has initiated us to come up with the complete revision of Indian Pandanaceae. This book is the original research work that provides not only updated taxonomic status of Indian screw pines but also their phylogenetic interrelationship in Indian and global context. The Pandanaceae family is distributed in two hotspots of India – Western Ghats and northeast Himalayan region and Andaman and Nicobar Islands. Our study has led in the addition of three new *Pandanus* species (two from Western Ghats and one from northeast Himalaya), merging of synonyms and totalling the species number to 14. Genus *Benstonea* with two species from Southern and Northeastern India; genus *Freycinetia* with two species, restricted in Andaman and Nicobar Islands. The species belonging to *Benstonea* and *Pandanus* have been described with respect to their biogeographical distribution, morphology, detailed species identification key and economic importance. The species have been assessed for their conservation status based on IUCN Red List Categories and Criteria (2001) revealing six species under threatened categories. The real outcome of the work is establishing phylogenetic relationship among the species based on the chloroplast DNA-based molecular phylogenetic approach. For the first time, the infrageneric classification for Indian *Pandanus* species is given. We hope that this book will be useful to taxonomists and phylogenetists working on screw pines and other related genera. It will also be a guideline for assessing evolutionary relationship among the taxa at different hierarchical levels using chloroplast DNA-based molecular phylogenetic approach.

Altafhusain Nadaf
Rahul Zanan

Acknowledgements

At the outset, we heartily express our gratitude to the most beneficent and most merciful Almighty for giving us inspiration and capacity to explore in His creation. We heartily thank our Hon'ble Vice-Chancellor Professor W. N. Gade for his encouragements and support. We would not have been able to explore throughout India without the financial assistance from the Board of College & University development (BCUD), University of Pune, Pune. We are indebted to the then BCUD Director Dr. Pandit Vidyasagar, Dean Faculty of Sciences and the then Officer on Special Duty (OSD) Dr. K. C. Mohite for providing us financial assistance. We are thankful to Prof. S. S. Deokule, Head and former Heads (Prof. B. B. Chaugule, Prof. V. R. Gunale) Department of Botany, University of Pune, Pune, for providing us necessary facilities.

We feel ourselves fortunate in getting associated with Dr. Martin W. Callmander, Assistant Curator, Missouri Botanical Garden, Saint-Louis, USA, and Conservatoire and Jardin botaniques de la Ville de Genève, Switzerland, who provided us his tireless support in the current study, carefully reviewing the new species. He has rewieved the taxonomic section of this book and suggested many improvement in several other sections. During the course of confirmation and authentication of Indian screw pines, he made available to us the photographs of Indian Pandanaceae herbaria deposited at the Kew Botanical Gardens (K), the Singapore Herbarium (SING), Herbarium Universitatis Florentinae (FI), the Royal Botanic Garden Edinburgh (RBGE) and Muséum National d'Histoire Naturelle de Paris (P) and sent us the photographs of type specimens. We are also indebted to him for writing foreword for this book. We would like to also thanks the Curators and collaborators of these institutions for helping Dr. Callmander to get the pictures that were primordial for this overview. We also put on record his help in tracing literature from various sources. We are thankful to the BSI, Calcutta and regional offices of Gangtok, Pune and Coimbatore for providing Pandanaceae herbaria and literature. We are thankful to Prof. S. R. Yadav, Shivaji University, Kolhapur, for introducing us to Dr. K. G. Bhat, Poornprajna College, Udupi, reviewing one of our manuscripts and encouraging us during this work. We are thankful to Dr. K. G. Bhat for his help in collecting *Pandanus* species from Karnataka state of India. Thanks are due to Dr. M. U. Sharief, National Orchidarium and Experimental Garden, BSI, Yercaud, for providing material on *P. leram*. We are thankful to Dr. P. G. Diwaker the then Joint Director, BSI, Pune, for providing leaf material of *P. leram*. We also thank the Head, Department of Physics, University of Pune for providing SEM facility.

We are thankful to the team of Springer India (Pvt.) Ltd. Life Sciences, New Delhi, for publishing this book. Finally, we thank our respective family members for their patience, support and blessings.

Authors

Contents

Chapter 1
World Pandanaceae: An overview

The Old World tropical screw pine family Pandanaceae represents palm-like trees with long, narrow, rigid, spirally arranged pineapple-like leaves and prop roots. The family has botanical interest because of its suggested pivotal position in the evolution of the monocotyledons (Meeuse 1965, 1966), unusual growth forms (Guillaumet 1973; Stone 1970; Tomlinson et al. 1970), and breeding systems (Hutchinson 1973; Rendle 1930; Stone 1968; Warburg 1900). The Pandanaceae, or screw pine family, is believed to be entirely dioecious (Cox 1984). The family contains five genera, *Freycinetia, Pandanus, Sararanga, Benstonea,* and *Martellidendron,* geographically distributed from seaside to high mountains (Wardah and Setyowati 2009; Callmander et al. 2012). Among the five genera, genus *Pandanus* has the broadest geographical distribution, occurring throughout the Old World tropics (West Africa eastward throughout tropical areas to the Pacific Islands), having approximately 500 species with tree and shrub habit. It is followed by *Freycinetia,* with a more limited range from Southeast Asia, the Pacific Islands, and Oceania (Ceylon eastward through the Malesian area into the Pacific and New Zealand), comprises nearly 200 species of lianas. Next to *Freycinetia,* genus *Benstonea* exhibits a coherent biogeographic distribution ranging from India to the South Pacific, with high species richness in South East Asia (Peninsular Malaysia and New Guinea). *Martellidendron* represents 6-foot-tall tree species endemic to Madagascar and the Seychelles islands. Finally, genus *Sararanga* is confined to the Philippines, New Guinea, and the Solomon Islands, with only two tall tree species (Callmander et al. 2003, 2012; Buerki et al. 2012; Stone 1983). The area-wise distribution at the global level is given in Fig. 1.1.

Geographical Distribution of Pandanaceae

The family is important in several regions wherein it has developed a high degree of endemism and contributes to the fundamental structure and physiognomy of the vegetation (Stone 1983). New Zealand represents only one species of *Freycinetia*; Thailand represents *Freycinetia* (two or three species) and over 24 *Pandanus* species; Malaya with 8 *Freycinetia* and about 50 *Pandanus* species; Borneo showed about 24 *Freycinetia* and more than 50 *Pandanus* species; the Philippine archipelago with about the same numbers of species of *Freycinetia* and *Pandanus* as Borneo; New Guinea representing about 60 *Freycinetia* species and more than 70 *Pandanus* species; New Caledonia with about 14 *Freycinetia* and 21 *Pandanus* species. Australia included four species of *Freycinetia* and about 20 *Pandanus* species. Fiji, Mauritius, Seychelles, Burma, the Himalayan foothills, Sumatra ,and East Africa represent some important secondary centers. Lord Howe Island, Christmas Island (Indian Ocean), and Hainan are several of the small regions having one or a few local endemic species. Sao Thomè Island (off Angola, West Africa) represents a single *Pandanus* species; Madagascar is a central area of endemism, representing about ca. 85 endemic *Pandanus* species and five species of

A. Nadaf and R. Zanan, *Indian Pandanaceae - an overview*,
DOI 10.1007/978-81-322-0753-5_1, © Springer India 2012

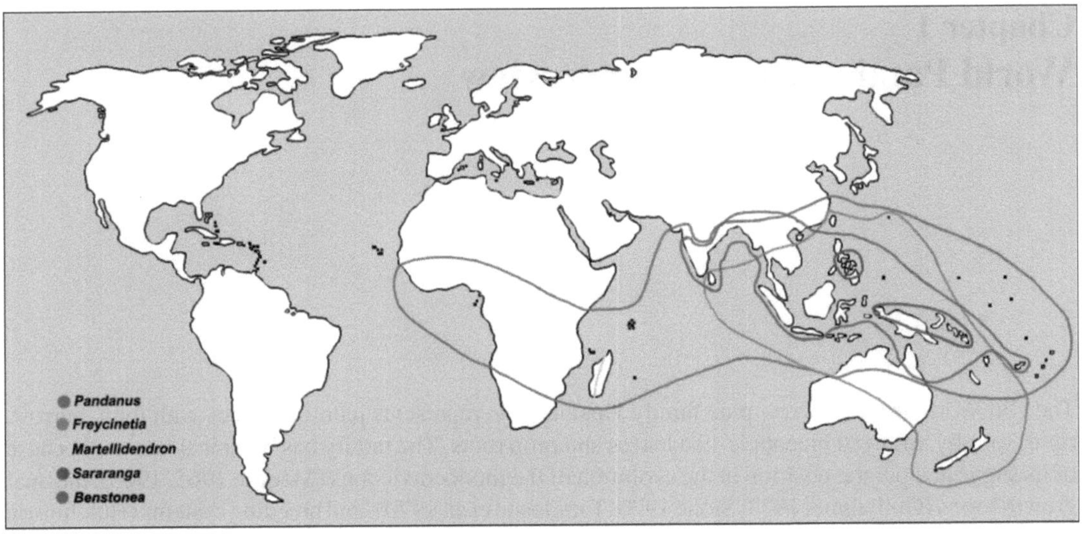

Fig. 1.1 World distribution of the family Pandanaceae

Table 1.1 Genus wise world distribution of Pandanaceae[a]

Genera	Distribution
Pandanus	Philippines, New Guinea, Africa, Solomon Islands, Madagascar, Malaya, Thailand, Borneo, New Caledonia, Australia, India, Sao Tomé Island
Freycinetia	Philippines, New Guinea, Solomon Islands, Malaya, Thailand, Borneo, New Caledonia, Australia, New Zealand
Martellidendron	Madagascar, Seychelles islands
Sararanga	Philippines, New Guinea, Solomon Islands
Benstonea	India, Sri Lanka, Thailand, Combodia, Laos, Viet-Nam, Philippines, Borneo, New Guinea, Solomon Islands, Fuji, Australia, Sulawesi, Java, Sumatra and Peninsular Malaysia (incl. Singapore)

[a]Based on Stone (1983) and Callmander et al. (2003, 2012)

Martellidendron (a sixth species of the latter genus is endemic to the granitic Seychelles Islands). The Solomon Islands represent about 23 species of *Freycinetia,* 28 species of *Pandanus*, a single species from *Sararanga* (Stone 1983; Callmander et al. 2003, 2011, 2012) (Table 1.1).

Morphology of Pandanaceae

The taxonomic position of Pandanaceae is as follows (APG III system 2009):

Kingdom: Plantae
Angiospearmae
Monocots
Order: Pandanales
Family: Pandanaceae R. Br.

The family Pandanaceae is evergreen trees, shrubs, and woody lianas. Stems are unbranched or branched with persistent annular leaf scars; aerial roots present or absent. Leaves are simple, spirally arranged at apex of stems, linear to lanceolate, leathery, glaucous, midrib and margin spinulose.

The inflorescences and infructescence are unique in the family. The flowers and fruits are unique in the family, being irregularly globular, many-seeded berries (Stone 1983). Inflorescence terminal, bracteate, spathes, white or colored, stamens numerous, filaments present, anthers basifixed; pollen grains often spinulose. Infructescence terminal, solitary, bracteates, irregularly globular with many-seeded berries, carpel free or aggregated, adjacent carpels forming one- to many-carpelled phalanges or clusters; ovary superior, one to several locular; ovules solitary to numerous; style absent or very short; stigmas one or more. Mesocarp is fibrous; endocarp membranous or evanescent; seeds are one to numerous, minute, oily, fleshy, or starchy (Sun and DeFilipps 2010; Stone 1983).

Pandanus species have a wide range of forms and size (Stone 1983), are tree-like and small shrub-like, and distinguished for having broad canopies and a moderate growth pattern. The long linear leaves originating at the stem appear screw-like and the edible fruits, to a great extent, resemble pine cones, hence the name "screw pine" (Badola et al. 2009). The male flowers are tiny, white, fragrant, and widely used for decoration; female flowers are edible (Thomson et al. 2006).

Freycinetia species are woody lianas; most species are rather smaller but three species are large plants and some are very small plants (Stone 1983).

Martellidendron is recognized by a 3-layered pollen exine, incomplete tectum, potentially bisexual flowers (presence of carpellodia and staminodia in the staminate and the pistillate flowers, respectively) and stigmas always in pairs forming a cross at the apex of the pileus and mesocarp extending between the seed locules (Callmander 2001; Callmander et al. 2003).

Sararanga species are arboreous, branched, leaves are 200 cm long or more with erect trunk, compound inflorescences, pendulous about 100 cm long (Stone 1983).

Benstonea is distinguished from other genera by its sharp spiniform style with stigmatic grooves consistently placed on the abaxial side, staminate flowers reduced to a single stamen or in triads that are free or very slightly joined at the base; epiphytic to mainly acaulescent shrubby habit (Callmander et al. 2012).

Origin of Pandanaceae

The major diversification of the flowering plants took place during the Early Cretaceous (Crane et al. 1995). Monocotyledons are among the major radiations of angiosperms, and they have been recognized as a group since studies of seed structure by Ray (1682, 1696, 1703). Many Cretaceous fossils of various groups of flowering plants have been discovered and described (Gandolfo et al. 2002; Crane et al. 1995), and the major branching pattern of the flowering plant phylogeny has been studied with the help of molecular cladistic analysis (Soltis et al. 1998; APG 1998). The combined study of fossil data and currently available molecular phylogenetic data provided the possibility of dating the entire phylogeny of flowering plants (Bremer 2000). Pandanales appeared to be well diversified by the late Cretaceous (Silva et al. 2011). The spinulose nature of the pollen in *Pandaniidites* and Pandanaceae is considered to be an apomorphy of the family (Hotton et al. 1994). The oldest fossil pollen of the family *Pandaniidites* were observed from 69 Mya (Miiller 1981). The macrofossil and palynological records of monocots from New Zealand showed that extinct Pandanaceae and Arecaceae were present in the Miocene (Pole 2007). Pandanaceae may date back to the Early Cretaceous (96 Mya) when Gondwanaland was still in the process of breaking up (Storey et al. 1995). The Pandanaceae are generally regarded as one of the lowest families of the monocotyledons (Campbell 1910). Dahlgren and Clifford (1982) regarded them as the sole representative of Pandanales in their morphological classification. *Sararanga* was considered the most primitive genus of Pandanaceae (Warburg 1900; Takhtajan 1969; North and Willis 1971; Stone 1972 and Dahlgren et al. 1985).

All five genera of Pandanaceae are paleotropical, but fossil genera *Pandaniidites* is known from the Americas, in Late Cretaceous and Paleocene sediments of North and South America (Miiller 1981; Jarzen 1983). The paleotropical distribution of the modem genera (many groups of flowering plants) indicates a Gondwanan origin dating from the later part of the Mesozoic (Callmander et al. 2003). The fossil record of Pandanaceae is small; hence, it is difficult to assess the center of origin. Bonde (1988) described *Pandanusocarpon* based on a fruit from the early Eocene of India, and Macphail et al. (1994) studied earliest Australian monocot fossils *Drytopollinites,* from the late Paleocene, assigned to *Freycinetia.* The poor fossil record of the pollen of *Freycinetia* suggest its recent origin (Callmander et al. 2003).

The presence of rich and endemic pandan flora (*Freycinetia, Sararanga*, and *Pandanus*) in New Guinea (Stone 1982) indicates a Gondwanan origin dating from the later part of the Mesozoic; family Pandanaceae migrated from eastern Gondwana through Asia and Australia through the Malay archipelago in Late Cretaceous (Audley-Charles 1987), and then established the *Martellidendron* linage in Madagascar and the Seychelles (Callmander et al. 2003). Africa represents the only genus *Pandanus* with large number of species in West Africa, indicative of its recent arrival and diversification in the country.

Contributors in Pandanaceae

Hendrik van Rheede (1636–1691)

Hendrik Adriaan van Rheede tot Drakenstein, a colonial administrator of the Dutch East India Company and naturalist, from 1670 to 1677, compiled the book *Hortus Malabaricus*, describing 740 plants in the Malabar region. In his time, the Malabar region referred to the stretch along the Western Ghats from Goa to Kanyakumari. The first volume of the *Hortus Malabaricus*, published in 1678, is a compendium of the plants of economic and medical value in the South Indian Malabar region. He continued his work for the next three decades and published 12 volumes in four languages: Latin, Sanskrit, Arabic, and Malayalam. In these volumes plants of the Malabar region are mentioned. Taxonomic treatment of Pandanaceae of India was presented by Rheede (1679) for the first time in *Hortus Malabaricus,* in which he described four species of *Pandanus* as Kaida, Kaida Taddi, Perin Kaida Taddi, and Kaida Tsjeria from the Malabar region.

Sydney Parkinson (1745–1771)

Sydney Parkinson was a Scottish Quaker, botanical illustrator, and natural history artist. He was employed by the British naturalist Joseph Banks to accompany him on James Cook's first voyage to the Pacific in 1768. Parkinson made nearly a thousand drawings on that voyage of plants and animals collected by Banks and the Swedish naturalist Daniel Solander. His work was finally published in 1988 by Alecto Historical Editions as *Banks' Florilegium* in 35 parts and has since been digitized by the Natural History Museum in London. He was the authority for the genus *Pandanus* and *Sararanga.*

Charles Gaudichaud-Beaupré (1789–1854)

Charles Gaudichaud-Beaupré was a French botanist associated with the French navigator Louis de Freycinet, who circumnavigated the world from 1817 to 1820 on the ships *Uranie* and *Physicienne*. He named the genus *Freycinetia* after Freycinet.

William Roxburgh (1751–1815)

William Roxburgh was a Scottish surgeon and botanist. He has been called the father of Indian botany. In the Indian context, he prepared the catalogue of the species of Calcutta Botanical Gardens – *Hortus Bengalensis* – and described some *Pandanus* species – *P. furcatus* Roxb., *P. foetidus* Roxb., and *P. amaryllifolius* Roxb.

Sir Joseph Dalton Hooker (1817–1911)

Sir Joseph Dalton Hooker was one of the greatest British botanists and explorers of the nineteenth century, and in 1893, he described *P. unguifer* from the Mungpoo hills in the Darjeeling district of West Bengal.

Harold St. John (1892–1991)

Harold St. John, after whom the St. John Plant Laboratory is named, was an eminent botanist and University of Harvard professor. He carried out extensive botanical expeditions throughout the Pacific Islands and opened the Pacific Islands to modern scientific discoveries. He is especially known for his work in discovering some 500 new species of *Pandanus* in the Pacific region. He revised genus *Pandanus* from the Pacific region in 44 parts and published in *Pacific Science*. His best known book is *List and Summary of the Flowering Plants of the Hawaiian Islands*.

Benjamin C. Stone (1933–1994)

Dr. Benjamin Stone worked extensively on tropical plant systematics, especially in the Old World region from the Indian Ocean to the Pacific, with special interest in Southeast Asia and Australasia and on the Pacific islands. He did field work in Madagascar, Sri Lanka, Malaya, Borneo, New Guinea, Melanesia, Micronesia, and Hawaii. He made great contributions in Pandanaceae, wherein he revised the whole family and developed an infrageneric classification in *Pandanus* that is still followed by classical and molecular taxonomists. The personal herbarium of Dr. Stone was deposited with the Academy of Natural Sciences of Philadelphia in the late 1980s and has long been recognized for the important material collected by him and others in Madagascar, Malaysia, New Caledonia, and other neotropical areas in the vicinity of Southeast Asia. It includes the large collection of Pandanus that represents Stone's work for over two decades. The Benjamin C. Stone Archive is maintained at the Botanical Institute of Texas (BRIT).

Martin W. Callmander (1975–)

Dr. Martin W. Callmander, a modern molecular taxonomist, is an assistant curator and technical advisor for the Conservation and Research Program in Antananarivo, Madagascar, Missouri Botanical Garden, USA and scientific collaborator for the Conservatoire et Jardin botaniques de la Ville de Genève, Switzerland. He is known for his numerous contributions to the systematics of the Malagasy flora in various families. He has been working on the family Pandanaceae since 1995. He recognized two new genera within the family: *Benstonea* and *Martellidendron* based on molecular phylogenies and morphology. He has contributed to the taxonomy of the family, especially in Madagascar and New Caledonia and started recently to work on South East Asia. A new species, *P. martinianus*, endemic to north-eastern India has been named after him by the authors as a tribute to his tirelessly support in the current revision of the Indian screwpines.

References

Angiosperm Phylogeny Group (APG) (1998) An ordinal classification for the families of flowering plants. Ann Missouri Bot Gard 85:531–553

Angiosperm Phylogeny Group (APG) (2009) An update of the Angiosperm Phylogeny Group classification for the orders and families of flowering plants: APG III. Bot J Linn Soc 161(2):105–121. doi:10.1111/j.1095-8339.2009.00996.x

Audley-Charles MG (1987) Dispersal of Gondwanaland: relevance to the evolution of the angiosperms. In: Whitmore TC (ed) Biogeographical evolution of the Malay archipelago. Cambridge University Press, Cambridge, pp 5–25

Badola HK, Pradhan BK, Subba S, Rai LK, Rai YK (2009) Sikkim Himalayan screw-pine, *Pandanus nepalensis*: a much neglected and underexplored NTFP. Non-Wood News 18:44–45

Bonde SD (1988) A new palm peduncle *Palmostroboxylon umariense* (Arecaceae) and a fruit *Pandanusocarpon umariense* (Pandanaceae) from the Deccan inter-trappean beds of India. In: Proceedings of the 3rd International Organization of Paleobotany conference, International Organization of Paleobotany, Melbourne, pp 59–65

Bremer K (2000) Early cretaceous lineages of monocot flowering plants. Proc Natl Acad Sci USA 97(9):4707–4711

Buerki S, Callmander MW, Devey DS, Chappell L, Gallaher T, Munzinger J, Haevermans T, Forest F (2012) Straightening out the screw-pines: a first step in understanding phylogenetic relationships within Pandanaceae. Taxon 61(5):1010–1020

Callmander MW (2001) Pandanus subg. Martellidendron (Pandanaceae) part II: revision of sect. Martellidendron Pic. Serm. in Madagascar. Bot J Linn Soc 137:353–374

Callmander MW, Chassot P, Küpfer P, LowryII PP (2003) Recognition of Martellidendron, a new genus of Pandanaceae, and its biogeographic implications. Taxon 52(4):747–762

Callmander MW, Laivao MO, Phillipson PB (2011) Pandanaceae specimens relocated from the University of Neuchâtel (NEU), with two new synonymies. Candollea 66(2):404–408

Callmander MW, Lowry II PP, Forest F, Devey DS, Beentje H, Buerki S (2012) Benstonea Callm. & Buerki (Pandanaceae): characterization, circumscription, and distribution of a new genus of screw-pines, with a synopsis of accepted species. Candollea 67(2):323–345

Campbell DH (1910) The embryo-sac of *Pandanus coronatus*. Bull Torrey Bot Club 37(6):293–295

Cox PA (1984) Chiropterophily and ornithophily in *Freycinetia* (Pandanaceae) in Samoa. Plant Syst Evol 144:277–290

Crane PR, Friis EM, Pedersen KR (1995) The origin and early diversification of angiosperms. Nature 374:27–33

Dahlgren RMT, Clifford HT (1982) The monocotyledons: a comparative study. Academic, London

Dahlgren RMT, Clifford HT, Yeo PF (1985) The families of the monocotyledons. Springer, Berlin

Gandolfo MA, Nixon KC, Crepet WL (2002) Triuridaceae fossil flowers from the Upper Cretaceous of New Jersey. Am J Bot 89:1940–1957

Guillaumet J (1973) Formes et developpement des Pandanus Malgaches. Webbia 28:495–519

Hotton CLK, Leffingwell HA, Skvarla J (1994) Pollen ultrastructure of Pandanaceae and the fossil genus Pandaniidites. In: Kurmann MH, Doyle JA (eds) Ultrastructure of fossil spores and pollen. Royal Botanic Gardens, Kew, pp 173–191

Hutchinson J (1973) The families of flowering plants. Oxford University Press, London

Jarzen DM (1983) The fossil pollen record of the Pandanaceae. Gard Bull Singapore 36:163–175

Macphail MK, Alley NF, Truswell EM, Sluiter IRK (1994) Early tertiary vegetation: evidence from spores and pollen. In: Hill RS (ed) History of the Australian vegetation: Cretaceous to recent. Cambridge University Press, Cambridge, pp 189–261

Meeuse ADJ (1965) Angiosperms-past and present: phylogenetic botany and interpretative floral morphology of the flowering plants. In: Vira R, Chandra L (eds) Advancing frontiers of plant sciences, vol 11. Impex, New Delhi, pp 170–182

Meeuse A (1966) Fundamentals of phytomorphology: 187-189, 206-207. Ronald Press, New York

Miiller J (1981) Fossil pollen records of extant angiosperms. Bot Rev 47:1–246

North CA, Willis AJ (1971) Contributions to the anatomy of *Sararanga* (Pandanaceae). Bot J Linn Soc 64:411–421

Pole M (2007) Monocot macrofossils from the Miocene of southern New Zealand. Palaeontol Electron 10(3):1–21

Ray J (1682) Methodus plantarum nova. Faithorne and Kersey, London

Ray J (1696) De variis plantarum methodis dissertatio brevis. Smith and Walford, London

Ray J (1703) Methodus plantarum emendata et aucta. Smith and Walford, London

Rendle A (1930) The classification of flowering plants. I. Gymnosperms and monocotyledons. Cambridge University Press, Cambridge

Silva RM, Santos DYAC, Salatino MLF, Motta LB, Cattai MB, Sasaki D, Lovo J, Pita PB, Rocini C, Rodrigues CDN, Zarrei M, Chase MW (2011) Five vicarious genera from Gondwana: the Velloziaceae as shown by molecules and morphology. Ann Bot 108:87–102. doi:10.1093/aob/mcr107

Soltis DE, Soltis PS, Mort ME, Chase MW, Savolainen V, Hoot SB, Morton CM (1998) Inferring complex phylogenies using parsimony: an empirical approach using three large DNA data sets for angiosperms. Syst Biol 47:32–42

Stone BC (1968) Morphological studies in the Pandanaceae. I. Staminodia and pistillodia of Pandanus and their hypothetical significance. Phytomorphology 18(4):498–509

Stone BC (1970) Malayan climbing *Pandans* the genus *Freycinetia* in Malaya. Malay Nat J 23:84–91

Stone BC (1972) A reconsideration of the evolutionary status of the family Pandanaceae and its significance in monocotyledon phylogeny. Quart Rev Biol 47:34–45

Stone BC (1982) New Guinea Pandanaceae: first approach to ecology and biogeography. In: Gressitt JL (ed) Biogeography and ecology of New Guinea, vol 1. Dr W. Junk, The Hague, pp 401–436

Stone BC (1983) A guide to collecting Pandanaceae (*Pandanus, Freycinetia, and Sararanga*). Ann Missouri Bot Gard 70(1):137–145

Storey M, Mahoney JJ, Saunders AD, Duncan RA, Kelley SP, Coffin MF (1995) Timing of hot spot-related volcanism and the break up of Madagascar and India. Science 267:852–855

Sun K, DeFilipps RA (2010) Pandanaceae. In: Zhengyi W, Raven PH, Deyuan H (eds) Flora of China, vol 23. Science press/MBG press, Beijing, pp 127–130

Takhtajan A (1969) Flowering plants. Origin and dispersal. Olivier & Boyd, Edinburgh

Thomson LAJ, Englberger L, Guarino L, Thaman RR, Elevitch C (2006) *Pandanus tectorius* (pandanus), ver. 1.1. In: Elevitch CR (ed) Species profiles for Pacific Island agroforestry. Permanent Agriculture Resources (PAR), Hōlualoa. http://www.traditionaltree.org

Tomlinson P, Zimmerman M, Simpson P (1970) Dichotomous and pseudodichotomous branching of monocotyledonous trees. Phytomorphology 20(1):28–32

Warburg O (1900) Pandanaceae. In: Engler A (ed) Pflanzenreich III. Verlag von Wihem Engelmann, Leipzig, pp 11–97

Wardah, Setyowati FM (2009) Ethnobotanical study on the Genus *Pandanus* L. f. in certain areas in Java, Indonesia. Biodiversitas 10(3):146–150

Chapter 2
Indian Pandanaceae: Present Status and Need of Revision

Present Status

The Pandanaceae are represented in India by three genera: *Pandanus* Parkinson, *Benstonea* Callm & Buerki and *Freycinetia* Gaudich distributed in South and Northeast India and Andaman and Nicobar Islands (Nadaf et al. 2011; Zanan and Nadaf 2012a, b). Genus *Freycinetia* confirmed from Andaman and Nicobar Islands (Zanan 2012). Taxonomic treatment of Pandanaceae of India began with van Rheede (1679), who described in *Hortus Malabaricus* four species of *Pandanus,* viz., Kaida (*P. kaida* Kurz.), Kaida Taddi (*P. odorifer* (Forssk.) Kuntze.), Perin Kaida Taddi (*P. unipapillatus* Dennst.), and Kaida Tsjeria (*P. furcatus* Roxb.). Later, in 1893, Hooker described *P. unguifer* from Mungpoo hills in the Darjeeling district of West Bengal. St. John, in his studies on Indian Pandanaceae, reported in 1965 a new species, *P. assamensis* from Assam, followed by the reports of five new species in 1972 – *P. diversus, P. emarginatus, P. katatonos, P. nepalensis,* and *P. sikkimensis* – from the Northeastern part of India. In *Wealth of India* (1966), 36 species of *Pandanus* are mentioned; however, a detailed list is not given. Vartak (1981) reported an exotic cultivated species, *P. amaryllifolius* found throughout India. In 1989, Karthikeyan et al. confirmed 19 *Pandanus* species from India. This was followed by the report of *P. dubius* (Keller 2001) and rediscovery of *P. unguifer* from Mungpoo in the Darjeeling district of West Bengal (Srivastava 1997; Srivastava and Chaudhary 2007). In a detailed exploration and survey, we reported three new species: *P. palakkadensis* from Palakkad, Kerala state (Nadaf et al. 2011), *P. mangalorensis* from Mangalore, Karnataka state (Zanan and Nadaf 2012a), and *P. martinianus* from the foothills of West Siang district of Arunachal Pradesh and Dhimaji district of Assam (Zanan and Nadaf 2012b). Recently, Callmander et al. (2012) recognized new genus *Benstonea* and shifted the species belonging to subgenus Acrostigma under the *Benstonea.* Thus, the family is described by several taxonomists and the total number of species belonging to Pandanaceae has not been reported. Further, there is no information available on the subgeneric and sectional levels or the phylogenetic relationship among the Indian *Pandanus* species.

Need of Revision

Stone (1976) stated that most of the *Pandanus* species are highly variable in morphology, and taxonomic distinctions between species and subspecies are often uncertain. Many species are morphologically similar and often distinguished by apparently minor differences of fruit structure (Stone 1974). Our review of literature showed that the taxonomy of the Indian *Pandanus* species is presently

A. Nadaf and R. Zanan, *Indian Pandanaceae - an overview*,
DOI 10.1007/978-81-322-0753-5_2, © Springer India 2012

in an unsatisfactory state. Several species have never been critically compared and were reported under different synonyms, which has created confusion in describing the total number. Further, several species were reported based upon poor data for specimens and, unfortunately, some have also been incorrectly described. A detailed account is given below.

In regional floras, *P. unipapillatus* was correctly reported and described (Gamble 1957; Sharma et al. 1984; Henry et al. 1989; Karthikeyan et al. 1989; Bhat 1992, 1997, 2003) even if as often been called with its later synonym described in 1900 by Warburg as *P. canaranus* Warb. *P. odorifer* (Datar 2007) was reported by different synonyms, viz., *P. tectorius, P. odoratissimus,* and *P. fascicularis* (Parkinson 1972; Gamble 1957; Karthikeyan et al. 1989; Parin 1981; Fischer. 1978; Singh 1988; Sharma et al. 1984, 1996; Rao 1986; Bhat 2003). *P. leram* from the Andaman and Nicobar Islands was reported as *P. andamanensium* (Karthikeyan et al. 1989; Parkinson 1972). Even the morphologically distinct species from the Northeast region and South India were merged, e.g., *P. nepalensis* from the Northeast region was mistaken as *P. furcatus* (St. John 1972). Polunin and Stainton (1984) and Hajra et al. (1996) reported *P. furcatus* as a synonym of *P. nepalensis* from Northeast India. Similarly, in the Northeastern region, J.D. Hooker (1878) first described *P. unguifer* and later considered it as conspecific to *P. minor* Buch. Ham. (1893), which was subsequently followed by most taxonomists (Srivastava 1997). Warburg (1900) and Martelli (1913) rejected this species and reduced it to the synonymy of *P. minor*. St. John (1972) reported a new species, *P. sikkimensis,* from the same locality where Hooker had reported it as *P. unguifer* (Noltile 1994). Later, Hajra et al. (1996) again reduced *P. minor* to the synonymy of *P. unguifer*. However, our studies, which are based upon the type specimen from Herbarium Universitatis Florentinae (FI), University of Florence, Florence, Italy (Herbarium No. 2698), indicated that *P. minor* and *P. sikkimensis* are synonyms of *P. unguifer.* The earliest known species, *P. thwaitesii,* was published in different floras (Gamble 1957; Sharma et al. 1984; Henry et al. 1989; Karthikeyan et al. 1989; Bhat 1992), but Bhat (1997) reported that *P. thwaitesii* does not occur in India and reported it as *P. foetidus*. Our survey revealed the occurrence of these two species from South India (Zanan and Nadaf 2011). Based on recent molecular phylogenetic analyses on Pandanaceae, *P. foetidus* and *P. thwaitesii* from genus *Pandanus* subgenus Acrostigma are shifted under the genus *Benstonea* (Buerki et al. 2012; Callmander et al. 2012).

In considering these multiple discrepancies, the need for a revision of the whole family became clear; hence, through extensive surveys over a period of 6 years, the most updated and revised information about the Indian Pandanaceae members has been gathered.

In the present study, we have redescribed all Indian *Pandanus* and *Benstonea* species in detail. In previous reports, large numbers of characters are not described, and most species are described based only on fruiting characters only, because of the ephemeral nature of staminate inflorescence and the lack of sufficient characters in vegetative parts (leaves) (Kam 1971). Here, we have taken into consideration vegetative (morphological and micro-morphology), and reproductive (inflorescence and infructescence) characters for delimitation and identification of Indian *Pandanus* and *Benstonea* species. The diversity of abaxial leaf epidermal papillae was observed by Solla (1884), Tomlinson (1965), and Kam (1971). Kam (1971) stated that distribution, size, and shape of papillae can be used for systematic purposes. In the present taxonomic treatment, we have used these micro-morphological characters as a preliminary tool for identification and delimitation of Indian *Pandanus* and *Benstonea* species. The Indian *Pandanus* and *Benstonea* species are revised and synonym species are merged to determine an exact number of species. Further, the species are assessed at subgeneric and sectional level to confirm their position in infrageneric classification. For the first time, a phylogenetic relationship between Indian *Pandanus* and *Benstonea* species is discussed using morphological characters and chloroplast DNA sequences.

References

Anonymous (1966) The Wealth of India. A dictionary of Indian raw material and industrial products, vol VII. Council of Scientific and Industrial Research, New Delhi, p 216

Bhat KG (1992) Studies on screw-pines (Pandanus) of Karnataka: new records. Indian J Forest 15:359–360

Bhat KG (1997) Does *Pandanus thwaitesii* Mart. occur in India? Rheedea 7(2):87–88

Bhat KG (2003) Flora of Udupi. Indian Naturalist, Udupi, pp 670–675

Buerki S, Callmander MW, Devey DS, Chappell L, Gallaher T, Munzinger J, Haevermans T, Forest F (2012) Straightening out the screw-pines: a first step in understanding phylogenetic relationships within Pandanaceae. Taxon 61(5):1010–1020

Callmander MW, Lowry II PP, Forest F, Devey DS, Beentje H, Buerki S (2012) Benstonea Callm. & Buerki (Pandanaceae): characterization, circumscription, and distribution of a new genus of screw-pines, with a synopsis of accepted species. Candollea 67(2):323–345

Datar MN (2007) An assessment of floristic diversity of Molem national park, Goa, vol II. Ph.D. dissertation, University of Pune, India, p 444

Fischer CEC (1978) Flora of the Anamalai hills. Bishen Singh Mahendra Pal Singh, Dehra Dun, p 186

Gamble JS (1957) Flora of the Presidency of Madras. Botanical Survey of India, Calcutta, pp 1094–1095

Hajra PK, Verma DM, Bandyopadhya S (1996) Flora of India, Series 2, Flora of Sikkim, vol I (monocot.). Botanical Survey of India, Calcutta

Henry AN, Chithra V, Balakrishnan AN (1989) Flora of Tamilnadu, vol III, Series 1, Analysis. Botanical Survey of India, Southern Circle, Coimbatore, p 54

Hooker JD (1893) Flora of British India, vol VI. L. Reeve & Co. Ltd, London, pp 483–488

Hooker JD (1878) Curtis's Botanical Magazine *Pandanus unguifer* – Native of Northern Bengal, vol XXXIV, L Reeve and Co., London, p 6347

Kam YK (1971) Morphological studies in Pandanaceae III. Comparative systematic foliar anatomy of Malayan Pandanus. Bot J Linn Soc 64:315–351

Karthikeyan SK, Jain SK, Nayar MP, Sanjappa M (1989) Flora of India, Series 4, Florae Indicae Enumeratio: Monocotyledonae. Botanical Survey of India, Calcutta, pp 177–178

Keller J (2001) Pandanaceae. In: Hanelt P, Institute of Plant Genetics and Crop Plant Research (eds) Mansfield's encyclopedia of agricultural and horticultural crops, vol V. Springer, Berlin, pp 2816–2824

Martelli U (1913) Enumerazione delle "Pandanaceae". Webbia; Raccolta de Scritti Botanici 4:5–105

Nadaf AB, Zanan RL, Wakte KV (2011) A new endemic species of Pandanaceae from India: *Pandanus palakkadensis*. Kew Bulletin 66:1–5. doi:10.1007/s12225-011-9265-0

Noltile HJ (1994) Flora of Bhutan, vol III, Part 1. Royal Botanic Garden, Edinburgh, pp 431–434

Parin D (1981) Bengal plants, vol II, Apocynaceae – Selaginellaceae. Bishen Singh Mahendra Pal Singh, Dehra Dun, pp 1100–1101

Parkinson CEP (1972) A forest flora of the Andaman Islands. Bishen Singh Mahendra Pal Singh, Dehra Dun, p 260

Polunin O, Stainton A (1984) Flowers of the Himalaya. Oxford University Press, New Delhi, p 142

Rao RS (1986) Flora of India, Series 2, Flora of Goa Diu Daman and Nagarhaveli. Botanical Survey of India, Calcutta, pp 451–452

Sharma BD, Singh NP, Raghavan RS, Deshpande UR (1984) Flora of India, Series 2, Flora of Karnataka analysis. Botanical Survey of India, Calcutta, pp 295–296

Sharma BD, Karthikeyan S, Singh NP (1996) Flora of Maharashtra State, Monocotyledons. Botanical Survey of India, Calcutta, p 206

Singh NP (1988) Flora of Eastern Karnataka, vol II. Mittal Publication, Delhi, pp 620–621

Solla RF (1884) Contribuzione allo studio degli stomi delle Pandanee. Nuov G bot ital 16:172–182

Srivastava RC (1997) *Pandanus unguifer* Hook. f. (Pandanaceae) – a little known species from Sikkim Himalaya. Natl Acad Sci Lett 20(3 and 4):44–45

Srivastava RC, Chaudhary RK (2007) Rediscovery of *Pandanus unguifer* Hook. f. (Pandanaceae): a RET species from its type locality after 75 years. Curr Sci 93(10):1325

St. John H (1965) Revision of the genus *Pandanus* Stickman. Part 20. New species from India and Thailand. Pac Sci 19:523–528

St. John H (1972) The Indian species of *Pandanus* (section Rykia). Bot Mag Tokyo 85:241–262

Stone BC (1974) Towards an improved infrageneric classification in *Pandanus* (Pandanaceae). Bot Jahrb Syst 94:459–540

Stone BC (1976) Pandanaceae. In: Saldanha CJ, Nicolson DH (eds) Flora of Hassan District, Karnataka, India. Amerind Pub. Co, Calcutta, pp 777–781

Tomlinson PB (1965) A study of stomatal structure in Pandanaceae. Pac Sci 19(1):38–54

Van Rheede HA (1679) Hortus Malabaricus, vol 2. Sumptibus Johannis van Someren, et Joannis van Dyck, Amsterdami

Vartak VD (1981) Note on Ambemohar-pat (*Pandanus amaryllifolius* Roxb.) from western India. J Bom Natl Hist Soc 78:196–198

Warburg O (1900) Pandanaceae. In: Engler A (ed) Pflanzenreich III, Verlag von Williem Engelmann, Leipzig, pp 11–97

Zanan RL, Nadaf AB (2011) Collection, characterization and olfactory evaluation of *Pandanus* species in Southern India. Plant Genet Resour 1–3. doi:10.1017/S1479262111000748

Zanan RL (2012) Molecular phylogeny of Indian Pandanaceae. Ph.D. dissertation, University of Pune

Zanan RL, Nadaf AB (2012a) *Pandanus mangalorensis*: a new species of Pandanaceae from southern India. Kew Bull 67:1–5. doi:10.1007/s12225-012-9366-4

Zanan RL, Nadaf AB (2012b) *Pandanus martinianus* (Pandanaceae), a new species endemic from North Eastern region of India. Phytotexa 73:1–7

Chapter 3
Biogeography of Indian Pandanaceae

Biogeography depicts the distribution and diversity of organisms in space and time (Cox and Moore 2005). Biogeography can be studied at ecological and historical levels to test hypotheses of species distribution and diversity (Lomolino et al. 2005). Ecological biogeography helps to understand distribution and diversity patterns based on the interactions between an organism and its physical and biotic environment (Huggett 2004). Historical biogeography determines the origin, dispersal, and extinction of tropical taxa (Crisci et al. 2003). Biogeography has been found useful for systematic conservation planning (Lourie and Vincent 2004; Whittaker et al. 2005). Systematically collected community data from different geographic regions provide an opportunity to evaluate existing biogeographic schemes and to compare how different community metrics reflect biogeographic variation seen in other groups of taxa (Shears et al. 2008). At high levels in the classification hierarchy, biogeographic data can be used to develop biologically meaningful planning units at reasonable spatial scales (Lourie and Vincent 2004; Whittaker et al. 2005).

Biogeographically, Pandanaceae is distributed in the Old World tropics (Callmander et al. 2003). All five genera of Pandanaceae are paleotropical, but fossil genera *Pandaniidites* is known from America, in Late Cretaceous and Paleocene sediments of North and South America (Miiller 1981; Jarzen 1983). The fossil records of Pandanaceae are less available; hence, it is difficult to assess the center of origin of the family (Callmander et al. 2003). *Pandanusocarpon,* from the early Eocene of India, was described based on a fruit (Bonde 1988), and one of the earliest Australian monocot fossils, from the Late Paleocene, is *Drytopollinites,* assigned to *Freycinetia* by Macphail et al. (1994). The fossil evidence of Pandanaceae suggests that the genus *Pandanus* appeared earlier than *Freycinetia,* which was confirmed through molecular analysis (Callmander et al. 2003). Stone (1982) reported that the New Guinea islands are rich in endemic Pandanaceae species (*Freycinetia, Sararanga*, and *Pandanus*). Based on this, Audley-Charles (1987) suggested a Gondwanan origin of Pandanaceae dating from the later part of the Mesozoic period, which further migrated from eastern Gondwana through Asia and Australia through the Malay Archipelago in the Late Cretaceous. Callmander et al. (2003) suggested the establishment of a recently discovered *Martellidendron* linage in Madagascar and the Seychelles. In Africa, only the genus *Pandanus* is reported, with a low level of morphological variation, indicating its recent spread associated with recolonization after the end of the drier climate regime of the Pleistocene (Callmander and Laivao 2002).

Biogeographic Distribution of Indian Pandanaceae

India is conveniently divided into ten major regions, based on geography, climate, and pattern of vegetation seen and on the communities of mammals, birds, reptiles, amphibia, insects, and other invertebrates that live there. These zones indicate a unique set of geo-physical and hydro-climatic condition as well as distinct geographical origin. They also have unique floral and faunal elements.

A. Nadaf and R. Zanan, *Indian Pandanaceae - an overview*,
DOI 10.1007/978-81-322-0753-5_3, © Springer India 2012

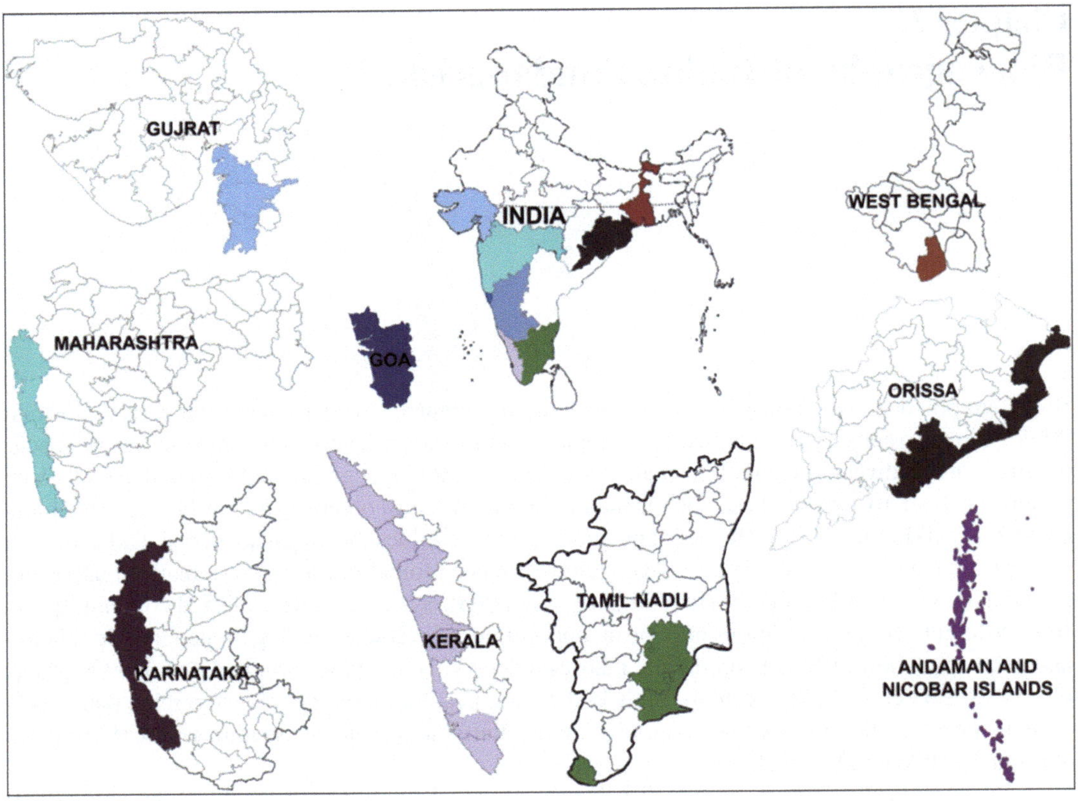

Fig. 3.1 Biogeographic distribution of *P. odorifer*

The Himalaya and Gangetic Plains are examples of two adjacent but extremely different zones (Rodgers et al. 2002).

In India, Pandanaceae members are distributed in tropical and subtropical forests of South and Northeast India and Andaman and Nicobar Islands. Based on our extensive survey of Pandanaceae throughout India, we have assigned Indian *Pandanus* and *Benstonea* species to these ten biogeographic zones and their provinces. The survey indicated that the *Pandanus* and *Benstonea* species are distributed in six biogeographic zones and ten provinces. Species-wise distribution patterns at world and Indian levels is detailed below.

1. ***Pandanus odorifer***: Distributed in South Asia, the Philippines, Polynesia, and tropical Australia (Thomson et al. 2006). In India, *P. odorifer* is distributed in two biogeographic zones: the Malabar Plains (province) of Western Ghats zone (Maharashtra, Goa, Karnataka, and Kerala) and the coastal zone (province: Long Western and Eastern Coastal belt with sandy beaches from Maharashtra, Goa, Karnataka, Kerala, Tamil Nadu, Andhra Pradesh, Orissa, and West Bengal). This species is also distributed in coastal regions of Andaman and Nicobar Islands (Fig. 3.1). It is one of the dominant species found in coastal zones of India.
2. ***Pandanus kaida***: Worldwide, this species is distributed from India to China. In India it is distributed in two different biogeographic zones: Malabar Plains (province) of Western Ghats (Goa, Karnataka, Kerala and Tamil Nadu) and Deccan Peninsular zone (province: South Deccan Peninsular from Tamil Nadu) (Fig. 3.2).
3. ***Pandanus furcatus***: Distributed in Pegu, Chittagong, and Malabar regions of India (Stone 1975). Biogeographically, *P. furcatus* is distributed in the Malabar Plains (province) of Western Ghats

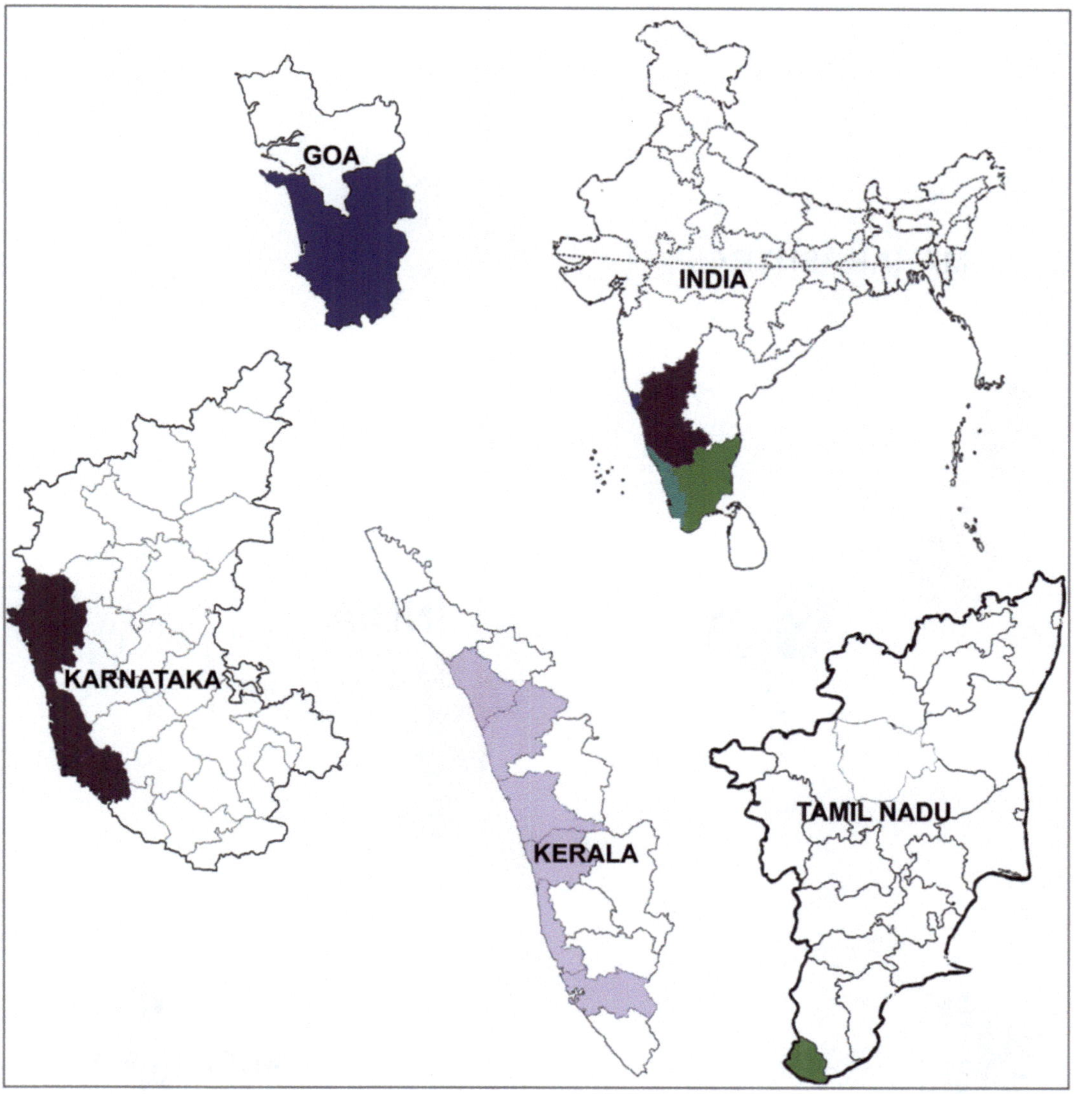

Fig. 3.2 Biogeographic distribution of *P. kaida*

zone (Maharashtra, Goa, Karnataka and Kerala). This species also distributed in the Andaman and Nicobar Islands (Fig. 3.3).

4. ***Pandanus unipapillatus***: Endemic to Southern India. Biogeographically, it is distributed in Western Ghats zone (Maharashtra, Goa, Karnataka, and Kerala) (province: Malabar Plains) (Fig. 3.4).
5. ***Pandanus palakkadensis***: Endemic to type locality from the Kerala state of Southern India (Nadaf et al. 2011). *P. palakkadensis* is observed in plain areas of the Palakkad plateau, Kerala under the Western Ghats (province: Malabar Plains) of biogeographic zones (Fig. 3.5).
6. ***Pandanus mangalorensis***: Endemic to type locality from the Karnataka state of Southern India (Zanan and Nadaf 2012a). *P. mangalorensis* is distributed in the Mangalore district of Karnataka, and this locality is biogeographically located under the Western Ghats zone (province: Malabar Plains) (Fig. 3.6).

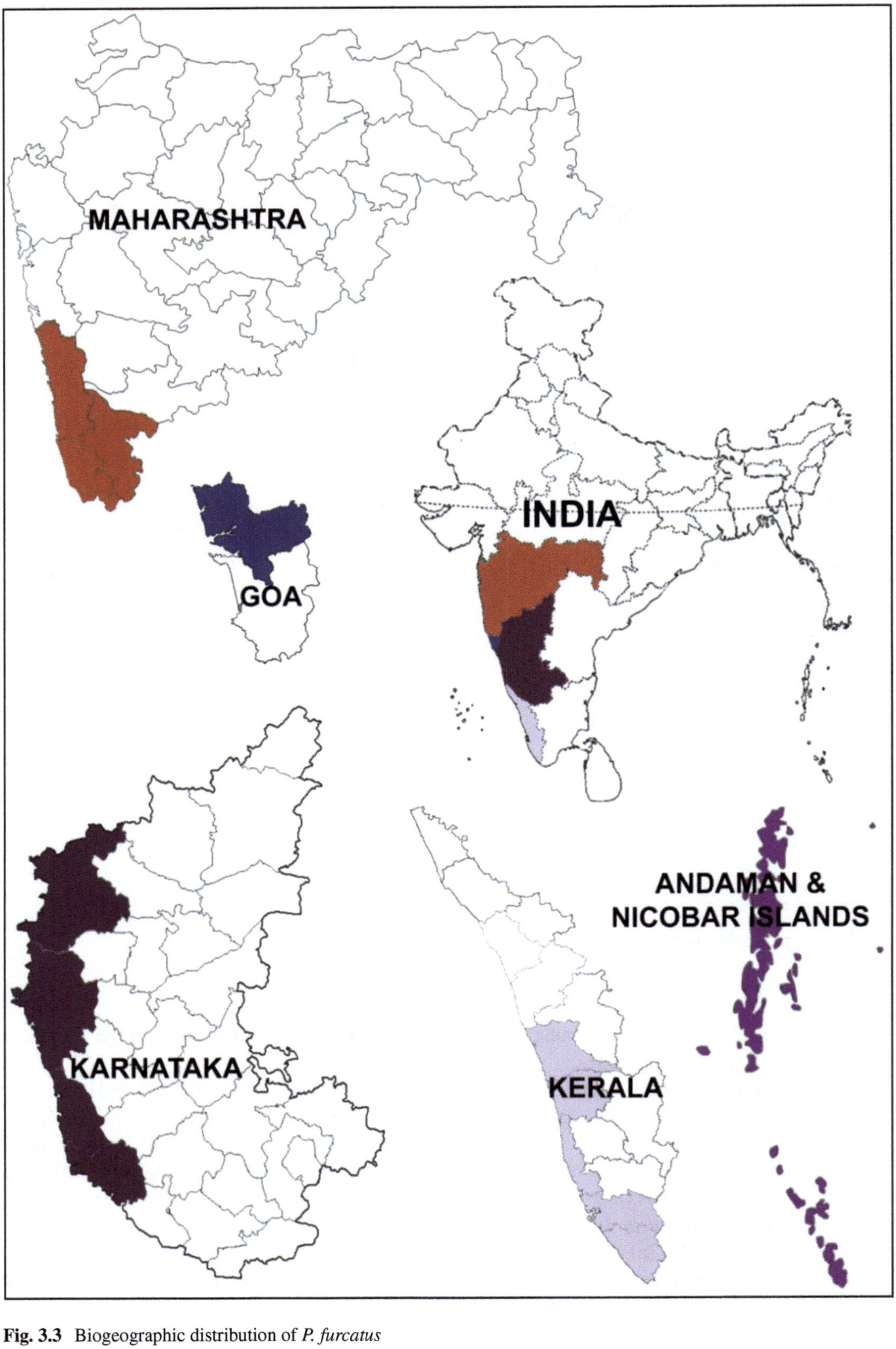

Fig. 3.3 Biogeographic distribution of *P. furcatus*

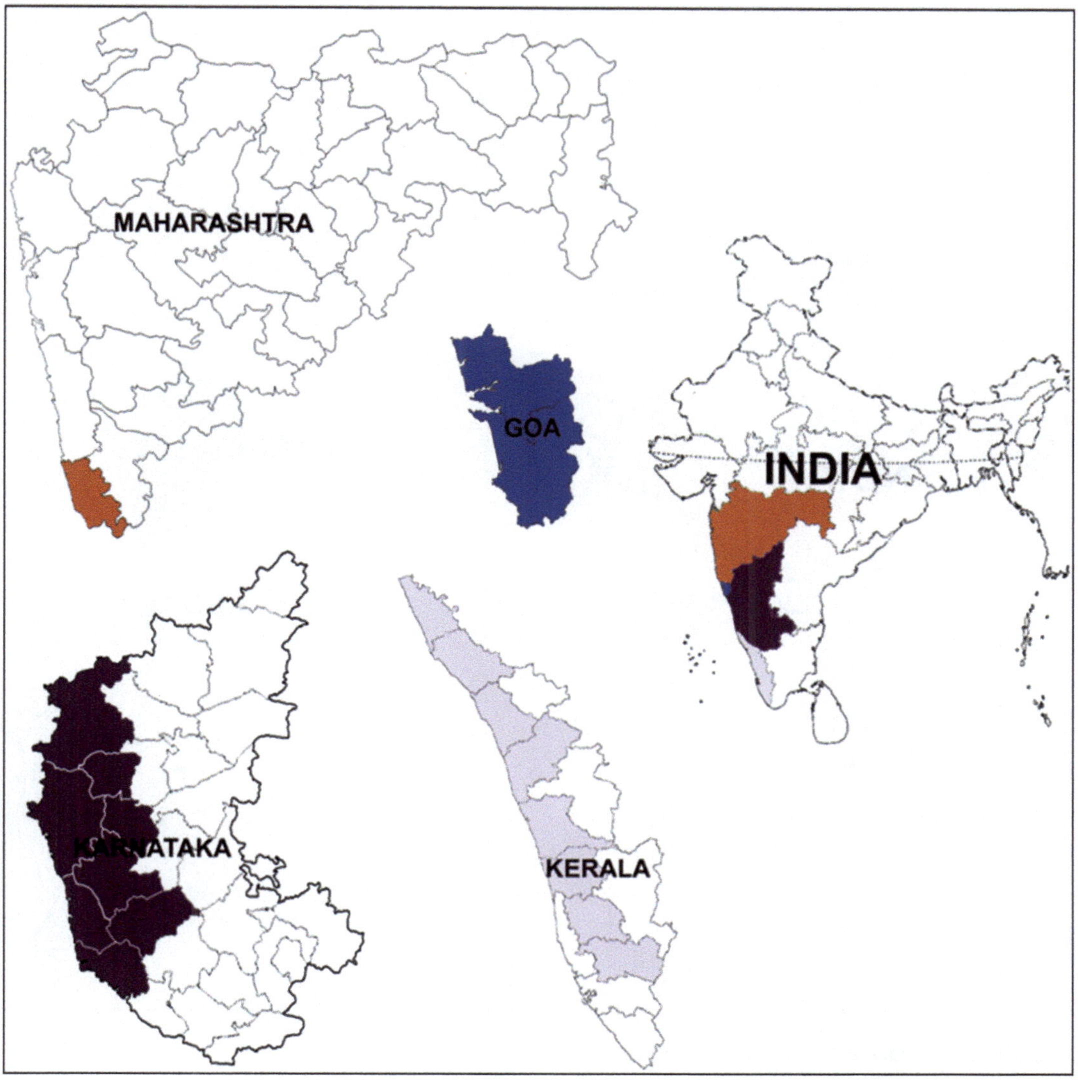

Fig. 3.4 Biogeographic distribution of *P. unipapillatus*

7. ***Pandanus dubius***: Distributed from Andaman and Nicobar Islands to West Pacific. It is indigenous to most of Malanesia and Polynesia; aboriginal introduction to Micronesia (Elevitch and Wilkinson 2000). In India, *P. dubius* is an exotic species, cultivated in Indian gardens of Kerala and Karnataka. Biogeographically, it is distributed in the Western Ghats zone of Karnataka (province: Malabar Plains) (Fig. 3.7).
8. ***Pandanus amaryllifolius***: Widely distributed in different tropical Peninsular countries, including Sri Lanka, Thailand, Southern India, Malaysia, Indonesia, Singapore, Vietnam, New Guinea, Taiwan, and the Philippines (Stone 1978; Ravindran and Balachandran 2005; Wakte et al. 2009). *P. amaryllifolius* originated in the region of the Moluccas Islands and has spread across the tropical region of Asia (Stone 1978). As stated elsewhere, the species was introduced in Botanical Garden of Calcutta (now Kolkatta) in 1798 (Stone 1978) and later spread across the coastal regions of India (Wakte 2010). In India, *P. amaryllifolius* is an exotic species, cultivated in

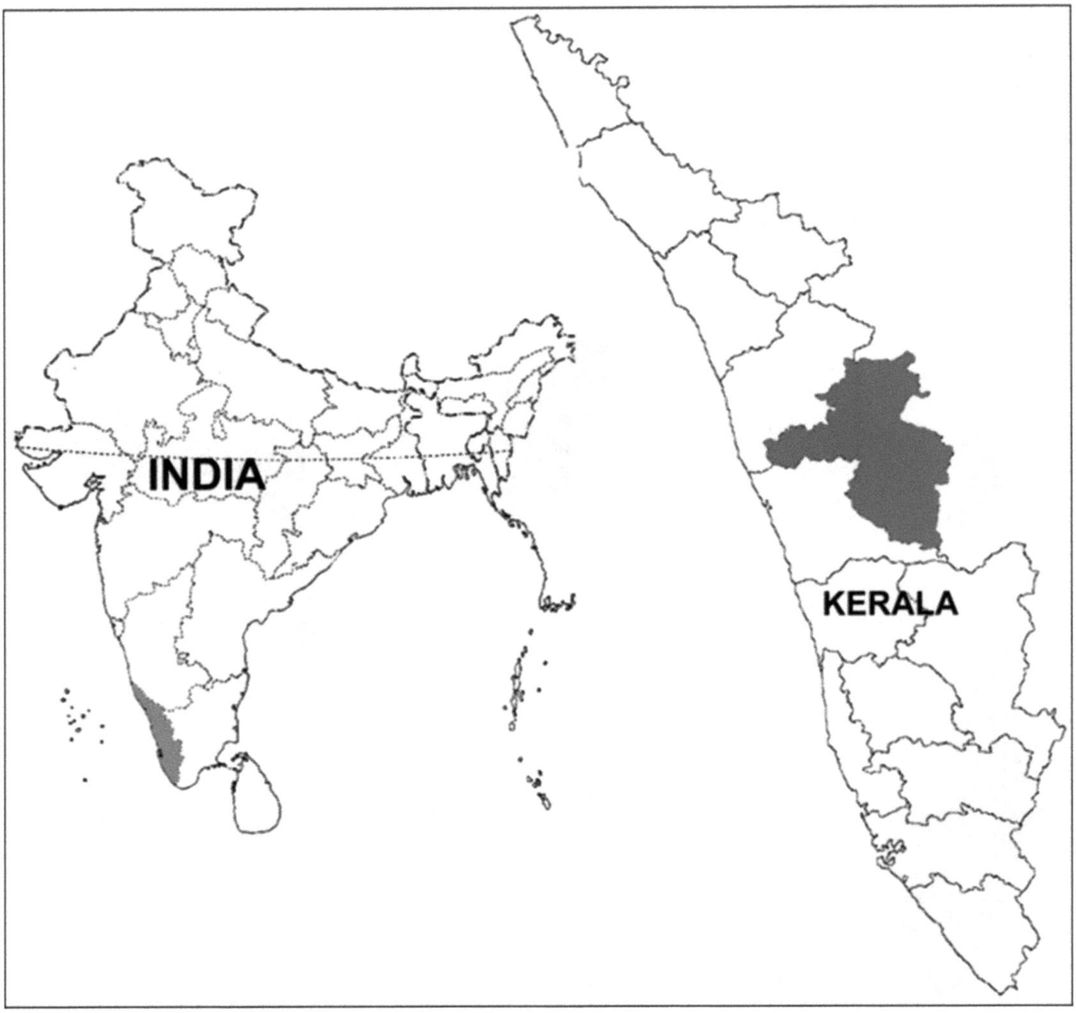

Fig. 3.5 Biogeographic distribution of *P. palakkadensis*

kitchen gardens (Wakte et al. 2009). Biogeographically, it is distributed in four different zones: the Western Ghats zone (province: Malabar Plains from Gujarat, Maharashtra, Goa, Karnataka, Kerala, Tamil Nadu), the Deccan Peninsular zone (provinces: Deccan South from Tamil Nadu and Eastern High Lands from Orissa), the Coastal zone (provinces: East coastal from Orissa and West Bengal), and the Gangetic Plains zone (provinces: Lower Gangetic Plains from West Bengal) (Fig. 3.8).

9. ***Pandanus leram***: Distributed from the Andaman and Nicobar Islands, South Sumatra, to West Java. *P. leram* is widely grown in the Andaman and Nicobar Islands and is biogeographically distributed in the Islands zone (provinces: Andamans and Nicobars) (Fig. 3.9). The single individual of the male species is cultivated in the Experimental Botanical Garden, Botanical Survey of India, Yercaud, Salem district, Tamil Nadu.
10. ***Pandanus nepalensis***: Distributed in Nepal, Assam, and probably in Bhutan (Stone 1975). In India, it is biogeographically distributed in two different biogeographic zones: the Gangetic Plain (province: Lower Gangetic Plains from Darjeeling district of upper West Bengal) and Himalaya (province: South Sikkim districts of Central Himalaya) (Fig. 3.10).

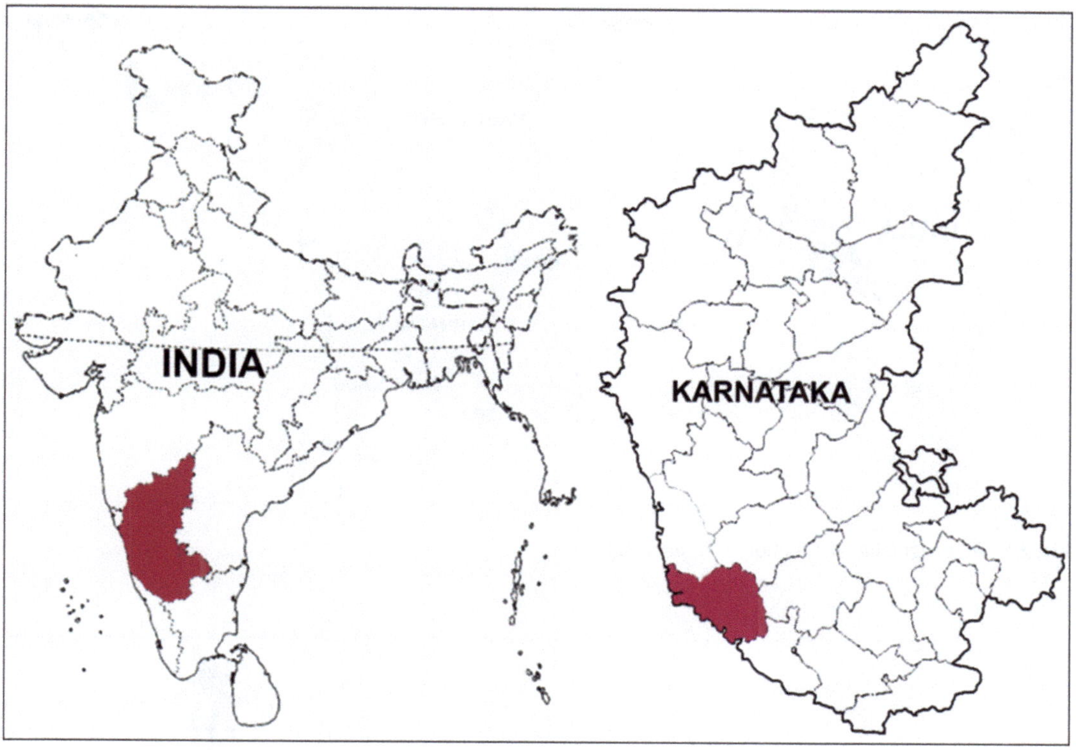

Fig. 3.6 Biogeographic distribution of *P. mangalorensis*

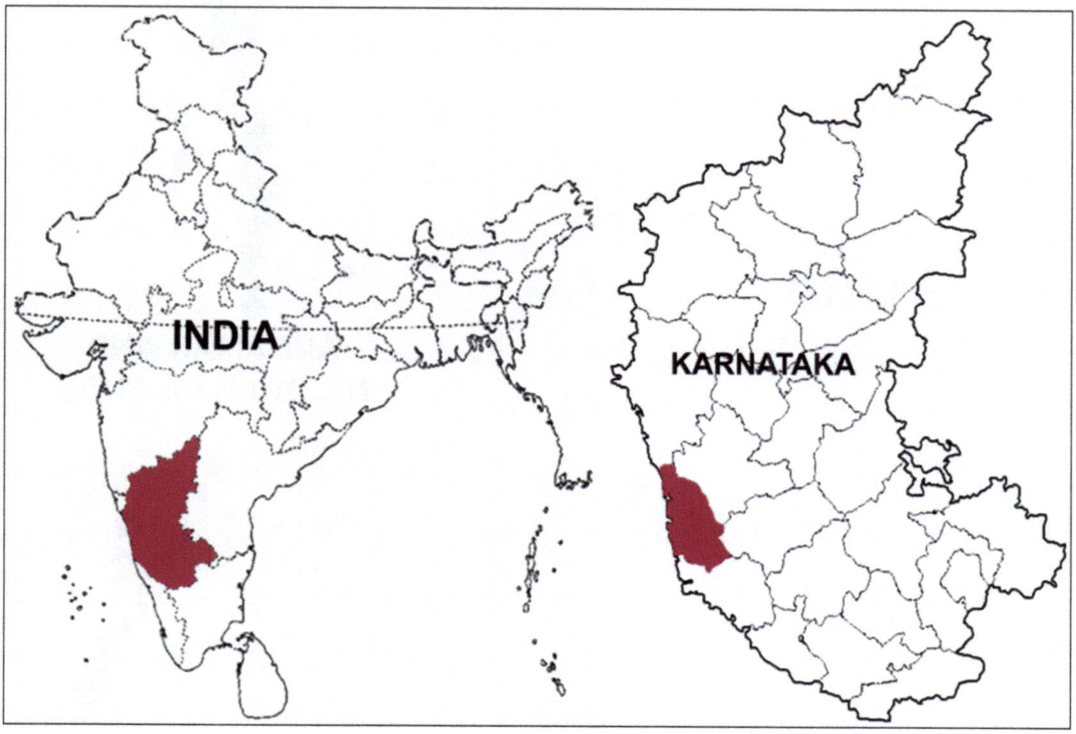

Fig. 3.7 Biogeographic distribution of *P. dubius*

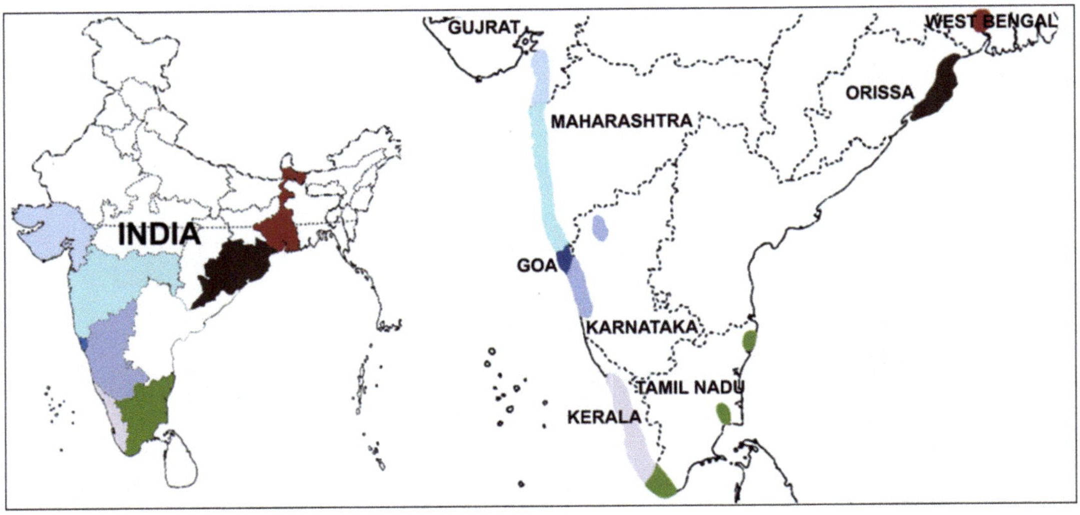

Fig. 3.8 Biogeographic distribution of *P. amaryllifolius*

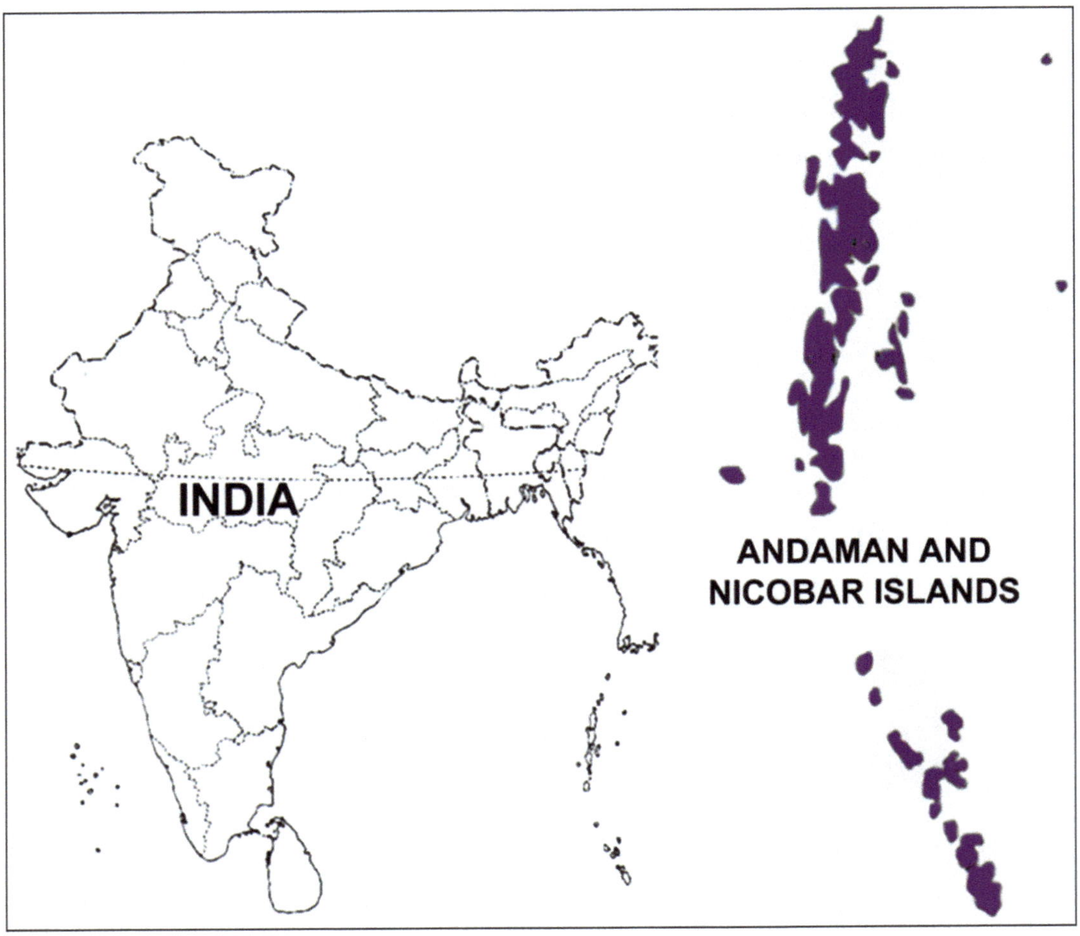

Fig. 3.9 Biogeographic distribution of *P. leram*

Fig. 3.10 Biogeographic distribution of *P. nepalensis*

Fig. 3.11 Biogeographic distribution of *P. emarginatus*

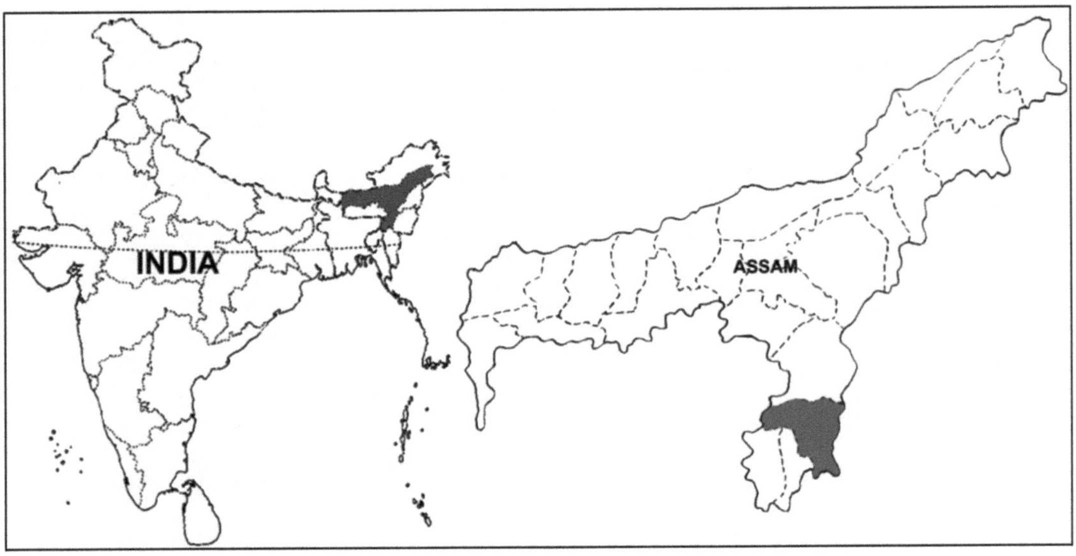

Fig. 3.12 Biogeographic distribution of *P. diversus*

11. ***Pandanus emarginatus***: Endemic to India (St. John 1972). Biogeographically, *P. emarginatus* is distributed in the Himalaya zone (province: Bhalukpong, West Kameng district of Arunachal Pradesh of East Himalaya) (Fig. 3.11).
12. ***Pandanus diversus***: Endemic to India (St. John 1972). In India, *P. diversus* is biogeographically distributed in the Northeast zone (province: Silchar district, Assam state of North East Hills) (Fig. 3.12).
13. ***Pandanus unguifer***: Distributed from Sikkim to Myanmar. In India, *P. unguifer* is biogeographically distributed in two biogeographic zones: Himalaya (province: South Sikkim districts of Central Himalaya) and Gangetic Plains (province: Darjeeling district of upper West Bengal from Lower Gangetic Plains) (Fig. 3.13).

Fig. 3.13 Biogeographic distribution of *P. unguifer*

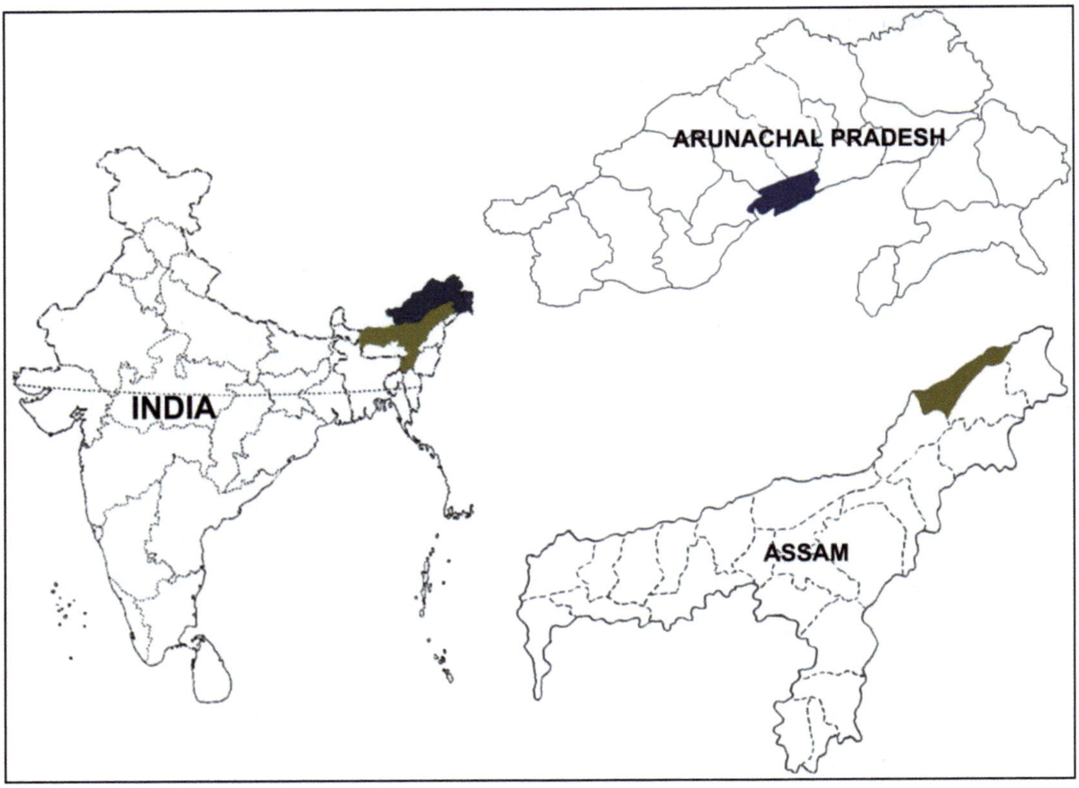

Fig. 3.14 Biogeographic distribution of *P. martinianus*

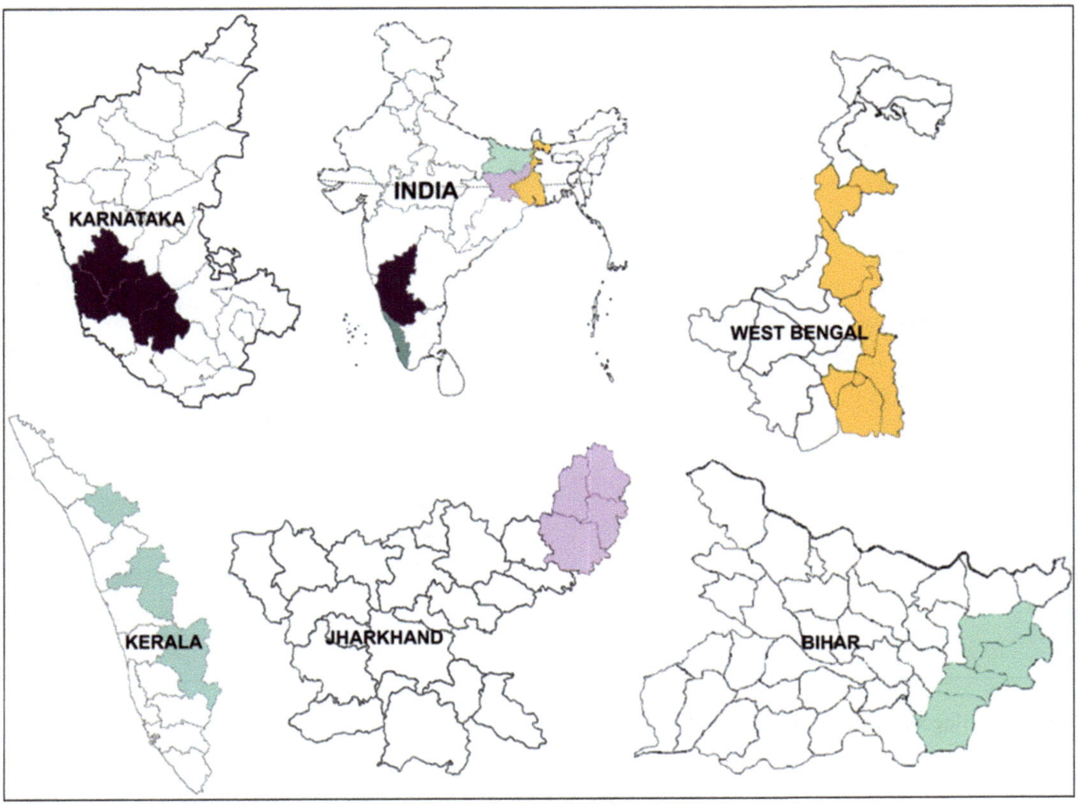

Fig. 3.15 Biogeographic distribution of *B. foetida*

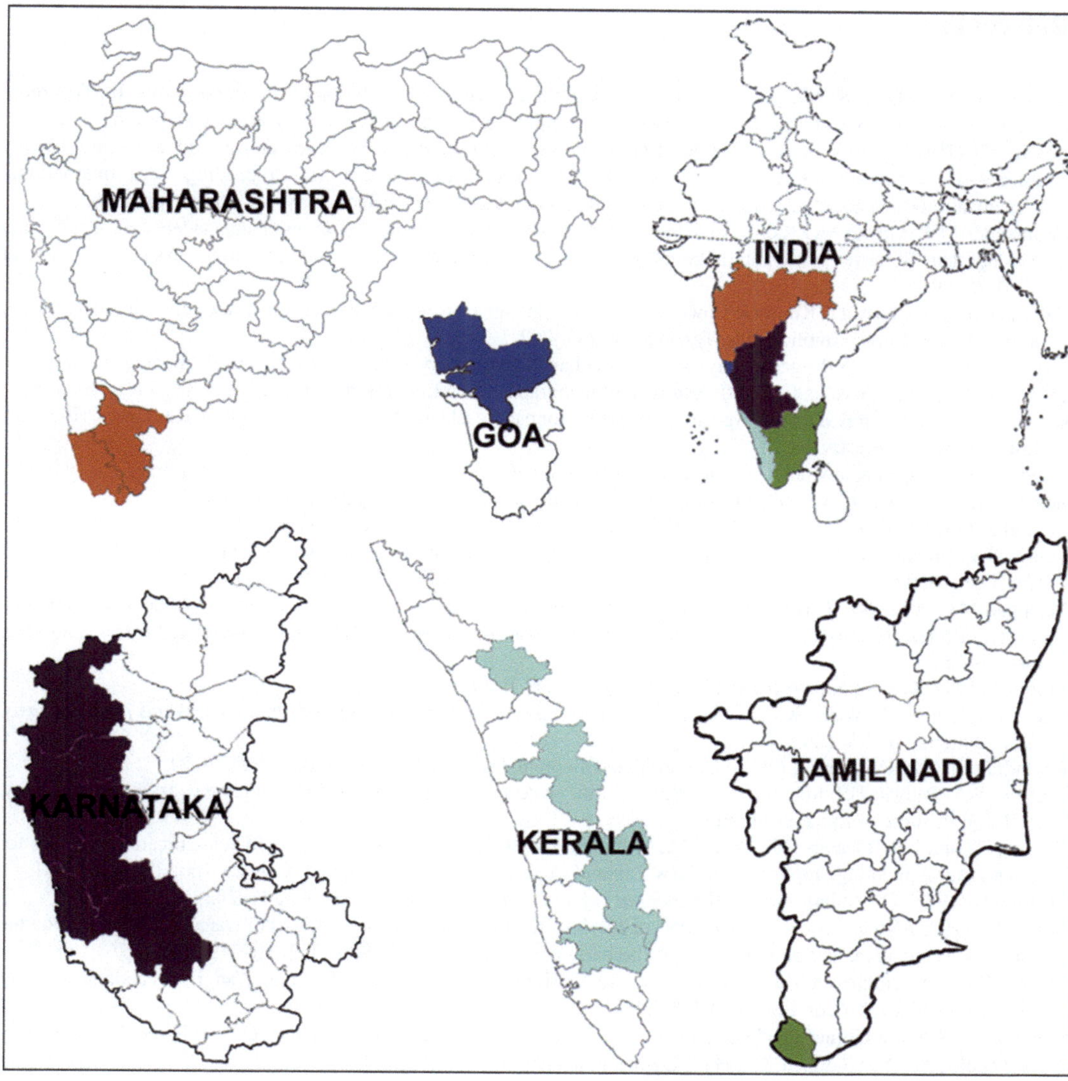

Fig. 3.16 Biogeographic distribution of *B. thwaitesii*

14. ***Pandanus martinianus***: Endemic to India (Zanan and Nadaf 2012b). Biogeographically it is distributed in two biogeographic zones: Himalaya (province: foothills of open forests in West Siang District of Arunachal Pradesh in East Himalaya) and the Northeast zone (province: Dhemaji District of Assam in Brahmaputra Valley) (Fig. 3.14).
15. ***Benstonea foetida***: Distributed in India and Burma. In India, *B. foetida* is distributed in three biogeographic zones: the Western Ghats zone (Karnataka and Kerala) (province: Western Ghats Mountains), the Coastal zone (province: East Coast of West Bengal), and Gangetic Plains (province: Lower Gangetic Plains of Bihar, Jharkhand and West Bengal) (Fig. 3.15).
16. ***Benstonea thwaitesii***: Distributed in South India and Sri Lanka. In India, *B. thwaitesii* is distributed in the Western Ghats zone (Maharashtra, Goa, Karnataka, Kerala, and Tamil Nadu) in two provinces: Western Ghats Mountains and Western Ghats Malabar Plains (Fig. 3.16).

References

Audley-Charles MG (1987) Dispersal of Gondwanaland: relevance to the evolution of the angiosperms. In: Whitmore TC (ed) Biogeographical evolution of the Malay Archipelago. Cambridge University Press, Cambridge, pp 5–25

Bonde SD (1988) A new palm peduncle *Palmostroboxylon umariense* (Arecaceae) and a fruit *Pandanusocarpon umariense* (Pandanaceae) from the Deccan Inter-trappean beds of India. In: Proceedings, 3rd International Organization of Paleobotany conference, International Organization of Paleobotany, Melbourne, pp 59–65

Callmander MW, Laivao MO (2002) Endémisme et biogéographie du genre *Pandanus* (Pandanaceae) en Afrique et à Madagascar. Bioterre Revue International des Sciences de Colloque international, Centre Suissu du 27–29 aout 2001:76–89

Callmander MW, Chassot P, Küpfer P, Lowry PP II (2003) Recognition of Martellidendron, a new genus of Pandanaceae, and its biogeographic implications. Taxon 52(4):747–762

Cox B, Moore PD (2005) Biogeography: an ecological and evolutionary approach. Blackwell Publishing, Malden

Crisci JV, Katinas L, Posadas P (2003) Historical biogeography: an introduction. Harvard University Press, Boston

Elevitch CR, Wilkinson KM (2000) Agroforestry guides for pacific islands. Permanent Agriculture Resources, Holualoa. http://www.agroforestry.net

Huggett RJ (2004) Fundamentals of biogeography. Routledge, New York

Jarzen DM (1983) The fossil pollen record of the Pandanaceae. Gard Bull Singapore 36:163–175

Lomolino MV, Riddle BR, Brown JH (2005) Biogeography. Sinauer Associates, Sunderland

Lourie SA, Vincent ACJ (2004) Using biogeography to help set priorities in marine conservation. Conserv Biol 18:1004–1020

Macphail MK, Alley NF, Truswell EM, Sluiter IRK (1994) Early tertiary vegetation: evidence from spores and pollen. In: Hill RS (ed) History of the Australian vegetation: cretaceous to recent. Cambridge University Press, Cambridge, pp 189–261

Miiller J (1981) Fossil pollen records of extant angiosperms. Bot Rev 47:1–246

Nadaf AB, Zanan RL, Wakte KV (2011) A new endemic species of Pandanaceae from India: *Pandanus palakkadensis*. Kew Bulletin 66:183–186

Ravindran PN, Balachandran I (2005) Underutilized medicinal species-III. Spice India 18(2):16–24

Rodgers WA, Panwar HS, Mathur VB (2002) Wildlife protected area network in India: a review, executive summary. Wildlife Institute of India, Dehradun

Shears NT, Smith F, Babcock RC, Duffy CA, Villouta E (2008) Evaluation of biogeographic classification schemes for conservation planning: application to New Zealand's coastal marine environment. Conserv Biol 22(2):467–481

St. John H (1972) The Indian species of *Pandanus* (section *Rykia*). Bot Mag Tokyo 85:241–262

Stone BC (1975) Notes on the Pandanaceae of Ceylon, with a review of the species cultivated in the royal botanic gardens, Peradeniya, and those found in its herbarium. Ceylon J Sci (Bio Sci) 11(2):109–125

Stone BC (1978) Studies in Malesian Pandanaceae XVII. On the taxonomy of 'Pandan Wangi': a Pandanus cultivar with scented leaves. Econ Bot 32(3):285–293

Stone BC (1982) New Guinea Pandanaceae: first approach to ecology and biogeography. In: Gressit JL (ed) Biogeography and ecology of New Guinea. The Hague, Boston, pp 401–436

Thomson LAJ, Englberger L, Guarino L, Thaman RR, Elevitch C (2006). *Pandanus tectorius* (pandanus), ver. 1.1. In: Elevitch CR (ed) Species profiles for pacific island agroforestry. Permanent Agriculture Resources (PAR), Hōlualoa. http://www.traditionaltree.org

Wakte KV (2010) Biochemical characterization of aroma volatiles and molecular genotyping in Indian *Pandanus amaryllifolius* Roxb. clones. Ph.D. dissertation, University of Pune

Wakte KV, Nadaf AB, Thengane RJ, Jawali N (2009) *Pandanus amaryllifolius* Roxb. cultivated as a spice in coastal regions of India. Genet Resour Crop Evol 56:735–740. doi:10.1007/s10722-009-9431-5

Whittaker RJ, Araujo MB, Paul J, Ladle RJ, Watson JEM, Willis KJ (2005) Conservation biogeography: assessment and prospect. Divers Distrib 11:3–23

Zanan RL, Nadaf AB (2012a) *Pandanus mangalorensis*: a new species of Pandanaceae from southern India. Kew Bull 67:1–5. doi:10.1007/s12225-012-9366-4

Zanan RL, Nadaf AB (2012b). *Pandanus martinianus* (Pandanaceae), a new species endemic from north eastern region of India. Phytotaxa 73:1–7

Chapter 4
Morphology of Indian Pandanaceae

The Indian Pandanaceae represents three genera viz., *Benstonea, Freycinetia* and *Pandanus*. The genus *Pandanus* is classified into 4 groups, 10 subgenera, and 59 sections (Stone 1974; Huynh 1991; Callmander et al. 2003). In relation to the infrageneric classification system (Stone 1974; Callmander et al. 2012), Indian *Pandanus* are represented in three subgenera: Rykia, Kurzia, and Pandanus. Table 4.1 depicts classification of Indian *Pandanus* species at the subgeneric level. The genus *Benstonea* and *Freycinetia* represents two species each. In the present study taxonomic description of genus *Pandanus* and *Benstones* has been provided.

Genus: *Pandanus* Parkinson

Pandanus Parkinson J. Voy. South Seas 46, 1773; Backer & Bakhuizen van den Brink. Fl. Java 3: 201, 1968; Stone, Biogeo. *Pandanus*. 69, 1975; Stone, Kew. Bull, 31: 57, 1976; Stone, Bot. J. Linn. Soc. 85: 133, 1982b; Brink & Jansen, Prosea 17: 197–204, 2003. Type species: *Pandanus tectorius* Parkinson – J. Voy. South Seas 46, 1773.

Trees or shrubs, usually erect or prostrate; stem simple to branched, smooth or armed with stout prickles, supported by prominent prop roots, prop roots smooth to armed with sharp prickles. Leaves linear or linear-lanceolate, glabrous, sheathing at the base, abruptly tapering to trigonous apex or gradually tapering to subulate apex, margins and midrib macroscopically spiny, apical ventral pleats absent to present or slightly two pleated. Large inflorescence, arranged in compound spikes, bracteates, every spike with a spatulate leafy bract, perianth absent. Stamens numerous, the filaments free or partly fused, anthers linear-oblong, basifixed. Infructescence arranged in terminal spikes or racemes; bracteates, ovate, sub-oblong to oblong or ellipsoid; carpels 1 – many in clusters or several-carpelled aggregates; phalanges or drupes cuneinate, clavate, narrowly oblanceolate or oblong; style subulate to bifurcate; stigma linear, subglobular-globular, cordate, lip-like, lanceolate or ellipsoid; pericarp thin, endocarp woody or bony; mesocarp sometimes hollow and basal mesocarp fibrous and fleshy; seeds ovoid or fusiform (Sun and DeFilipps 2010).

Subgenus: Rykia (de Vriese) B. C. Stone

Subgenus Rykia is polytypic and largest subgenus of genus *Pandanus*. India represents 4 sections with a total of 12 species. This distinct subgenus can be recognized by the combination of the following

In collaboration with Martin W. Callmander for the taxonomic treatment. Missouri Botanical Garden, St. Louis, U.S.A. and Conservatoire et Jardin botaniques de la ville de Genève, Switzerland.

A. Nadaf and R. Zanan, *Indian Pandanaceae - an overview*,
DOI 10.1007/978-81-322-0753-5_4, © Springer India 2012

Table 4.1 Classification of Indian *Pandanus* genus at sub generic level[a]

Genus	Group	Subgenus	Section	Subsection	Species
Pandanus Parkinson	Group I	Rykia (de Vriese) B. C. Stone	*Rykia* (de Vriese) Kurz	*Rykia* (de Vriese) Kurz	*P. furcatus* Roxb.
					P. unipapillatus Dennst.
					P. palakkadensis Nadaf, Zanan & Wakte
					P. mangalorensis Nadaf & Zanan
					P. nepalensis St. John.
					P. emarginatus St. John.
					P. diversus St. John.
					P. unguifer Hook. f.
					P. martinianus Nadaf & Zanan
			Hombronia (Gaudich.) Warb.	–	*P. dubius* Spreng.
					P. leram Jones *ex* Fontana
			Kaida St. John Ms.	–	*P. kaida* Kurz
		Kurzia B. C. stone	*Jeanneretia* (Gaudich.) B. C. Stone	–	*P. amaryllifolius* Roxb.
	Group III	Pandanus B. C. Stone, non St. J.	*Pandanus* (syn. Keura [Forssk] Warb.)	Pandanus (syn. Keura [Forssk] Warb.)	*P. odorifer* (Forssk.) Kuntze
Benstonea Callm. & Buerki					*B. thwaitesii* (Martelli) Callm. & Buerki
					B. foetida (Roxb.) Callm. & Buerki

[a]Based on Stone (1974); Callmandar et al. (2012). Based on our field visits and collections from throughout India and previously published literature species wise morphological account of Indian *Pandanaceae* is given below. In addition, micro-morphological details and species wise note is also given

characters: carpel free or connate into phalanges; style acute, deltoid-ovate, spiniform or bifurcate; stamens in phalanges, crowded at the apex of stemonophore. Leaf apex with ventral plates always unarmed.

Section: *Rykia* (de Vriese) Kurz; Subsection: Rykia (de Vriese) Kurz

This is a large section, well-developed in India and the surrounding regions. The *Rykia* section consists of seven species, known from the South and Northeastern regions of India. The species are morphologically similar, making identification difficult. In most cases, as is usual in the *Pandans*, the staminate plants remain unknown. The section Rykia, subsection Rykia can be recognized by the following combination of characters: carpels free fruits as one-seeded drupes; style spiniform, forked, simple or both types in one cephalium; cephalia solitary, endocarp longer than basal mesocarp; leaves with unarmed apical ventral plates.

Pandanus furcatus **Roxb.,** Hort. Bengal. 71, 1814; Fl. Ind. 3: 744, 1832; Kurz, Jour. Bot. Brit. and For. 5: 95, 1867; Solms, Linnaea 42: 13–14, 1878; Balf. F., Linn. Soc. Bot., Jour. 17: 47–48, 1878; Warb., Pflanzenreich IV, 9:75, fig. 21 A – C, 1900; Martelli, Webbia 4 (1): 15, 50, 96, 1913; 4 (2): t. 29, fig. 5–8, 1914; Miq., Fl. Ind. Bat. 3: 162, t. 37, 1855; Hook. f., Fl. Brit. India 6: 484, 1893;

Cooke, Fl. Pres. Bombay 3: 324, 1958 (Repr. ed.). *Rykia furcata* (Roxb.) de Vriese, K1. Akad. Vet., Verhandl. 203, 1854; de Vriese, Hook. Kew Jour. 6: 268, 1854. *Kaida Tsjerria* Rheede, Hort. Malab. 2: 7, t. 8, 1679.

Original diagnosis:
Roxburgh (1814): "shrubby." Drupes of the oblong compound fruit cuneate, crowned with an incurved polished, sharp, forked spine, nuts one-celled.

Kaida Tsjerria Rheede: "Est et alia species *Kaid* nascens in montosis. *Fructus* hujus (in quibus à prioribus speciebus differt) haut dissimiles fructibus *Tsjáka,* forma oblongo-rotundi et angustiores, in superficie bullulis seu oculis extuberantibus dotati, et oblongis, nigricantibus, rigidis spinis in vertice bullarum muniti, cortice viridi, ac odoris vegeti ac grati. Earundem cum caeteris est virium."

Expanded diagnosis:
A large shrub or small tree 5–8 m tall; stem erect, smooth, divaricately branched, branching near top, few prop roots at the base; leaves up to 130–180 cm long, 4–5 cm wide, margins and midrib spiny with three rows of whole leaf, margin with sharp prickles and midrib with curved prickles at the base, prickles yellow, on margin at the base 3 mm long with 8–10 mm distance, on midrib 2–3 mm long with 20–30 mm distance, at the middle part 7–8 spines with 2–3 mm long on midrib (per 10 cm), 15–16 spines along the margin, 2–3 mm long (per 10 cm) and crowded towards apex.

Inflorescence terminal, composite spike with 8–10 white bracts, variable in length, linear lanceolate, less fragrant, margin and midrib with prickles; staminate spike 8–12 cm long, cylindrical; stamen 6–10, androphore 8–10 mm long, stamenophore 4–5 mm long, stamens 4–5 mm long, anther 3–4 mm long, filament free, 0.5–1 mm long.

Infructescence terminal, bracteates with variable length, solitary, ellipsoid, 30–35 cm long, 18–22 cm diameter, drupes 35–45 mm long, 8–13 mm wide, hexagonal, compressed, pileus pyramidal, 8–12 mm long, smooth; style 3–4 mm long, almost at right angle, bony, yellowish, with bifid stigma, 1–1.5 mm long, brownish; endocarp abonic, central projection extended up to style base without shoulders, 15–17 mm long, 7–10 mm diameter, 1.5–2 mm thick with 14–20 mm long tip attached to pileus; mesocarp fibrous, apical 14–20 mm long, basal 13–15 m long; seeds 14–16 mm long, 3–4 mm diameter (Figs. 4.1 and 4.2).

Flowering: July to October, **Fruiting:** July to December.

Leaf micro-morphology: (Fig. 4.3)
In *P. furcatus* the abaxial epidermis is covered by papillae. The epidermal papillae are unbranched, without stack, size wise distinguishable as small triangular (5–6 μm in long) and large dome shaped (11–12 μm long). The dome-shaped papillae are taller than lateral subsidiary cell papillae. The terminal subsidiary cells papillae around stomata are branched and rod shaped (12–20 μm long), partially overarched on guard cells of stomata and other epidermal cells. Lateral subsidiary cell papillae 3–6, simple, globular, triangular and/or dome shaped, forming walls surrounding stomata. According to Tomlinson (1965) and Kam (1971), abaxial epidermal papillae of *P. furcatus* showed Class 4 (papillose neighboring and subsidiary cells) type of stomata. In *P. furcatus,* Tomlinson (1965) reported the presence or absence of papillae on terminal and lateral subsidiary cells.

Notes:
Pandanus furcatus is placed subgenus Rykia; it is endemic to the Malabar region only but botanists have considered it occurring all the way from India to Java (St. John 1972). We have sorted out this confusion and confirmed its location along the Malabar region only. A detailed review of literature is as follows. *Pandanus furcatus* was first described by van Rheede (1679) from the mountain region of Malabar, South India. In 1814, Roxburgh, based on Rheede's illustrations published the name as *P. furcatus* from Northeast India cited Kaida Tsjerria Rheede as a synonym (St. John 1972). To date, there is no

Fig. 4.1 *Pandanus furcatus*: (**a**) habit; (**b**) staminate inflorescence; (**c**) stamens; (**d**) syncarp; (**e**) stigmas in close view; (**f**) young syncarp

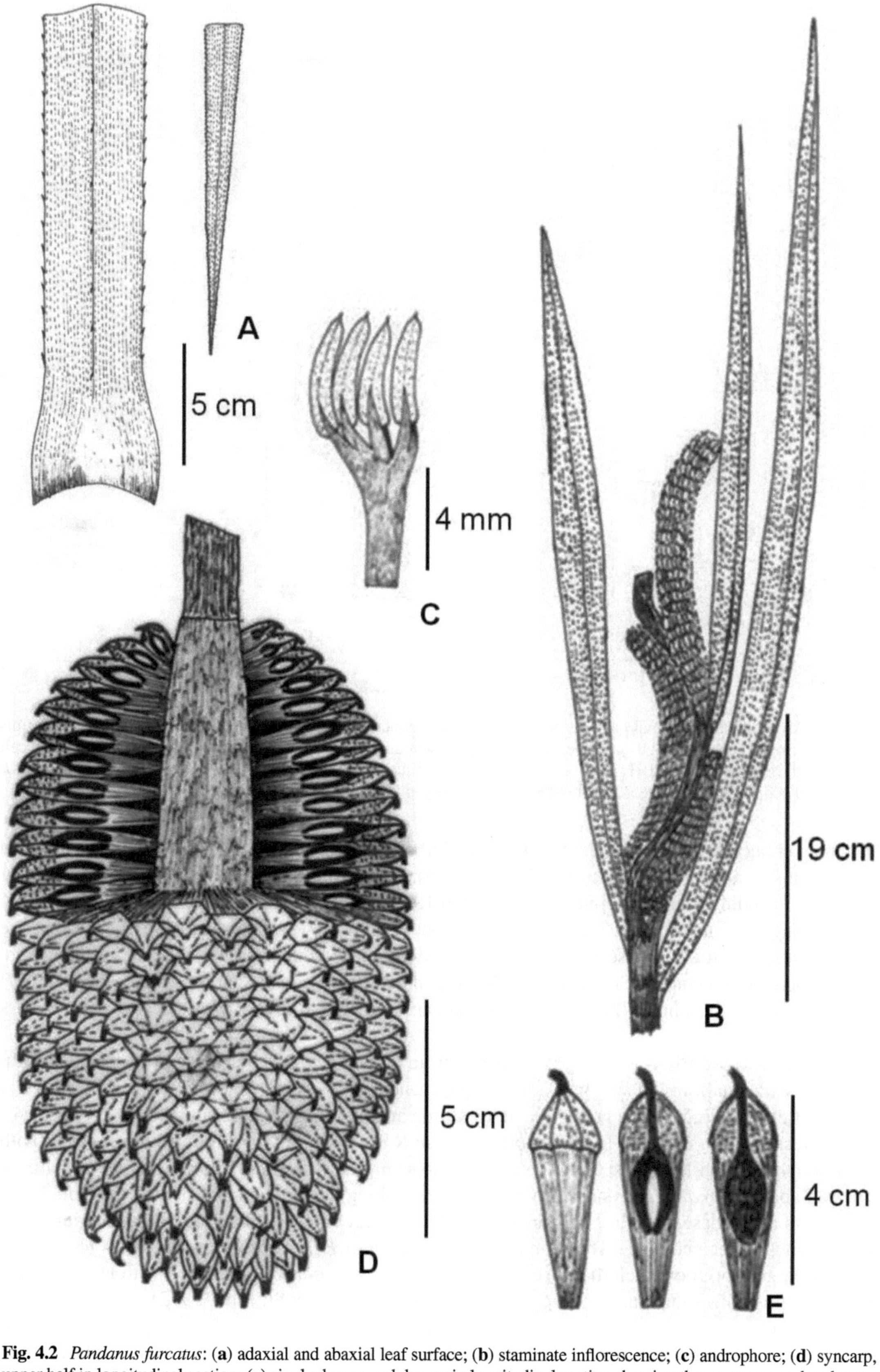

Fig. 4.2 *Pandanus furcatus*: (**a**) adaxial and abaxial leaf surface; (**b**) staminate inflorescence; (**c**) androphore; (**d**) syncarp, upper half in longitudinal section; (**e**) single drupes and drupes in longitudinal section showing the mesocarp and endocarp

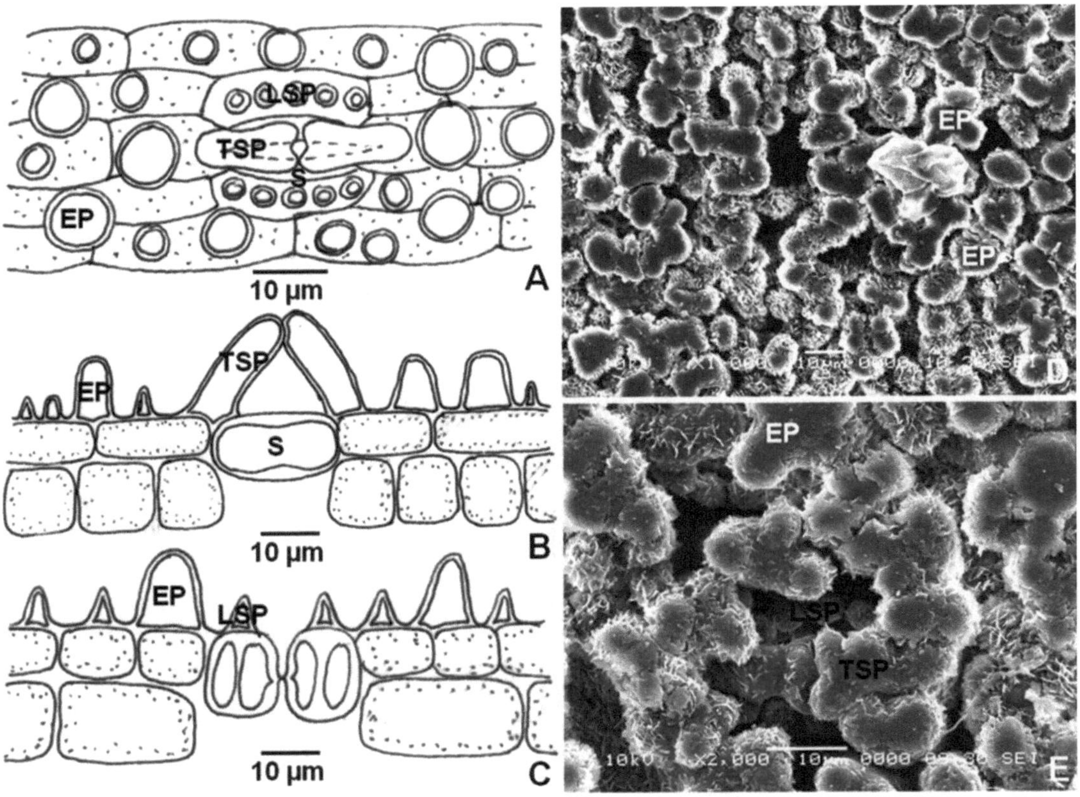

Fig. 4.3 Abaxial leaf epidermal papillae in *P. furcatus*: (**a**) surface view; (**b**) transverse section of abaxial epidermis showing terminal subsidiary cells papillae; (**c**) longitudinal section of abaxial epidermis showing lateral subsidiary cell papillae; (**d**) SEM of papillae (×1,000); (**e**) close view around stoma (×2,000) (*S* stomata, *EP* epidermal papillae, *TSP* terminal subsidiary cell papillae, *LSP* lateral subsidiary cell papillae)

matching collection available at the BSI Kolkata Herbarium from Malabar including the one described by Roxburgh. Later, the existence of *P. furcatus* in Northeast India was accepted and followed by other botanists, including Stone (St. John 1972). Prior to 1881, Humblot collected some seeds of *P. furcatus* from Madagascar and propagated in Kew Gardens, but St. John (1972) stated that there is no such native species known from Madagascar. After maturity, they produced male flowers and fruits, which were again described in detail by Stapf, wherein he determined and observed that the species was really *P. furcatus* of India (St. John 1972). Today, the only description available for *P. furcatus* is based on Stapf's description.

According to van Rheede, this species differs from others in having narrow, oblong rounded fruit, provided with small, round swelling or bulging out and oblong eyes; protected with a rigid spike at the top of the swelling, with a green rind and a strong and pleasing smell (Manilal and Suresh 1984). The collection from Malabar India (reported by van Rheede and by our group) did not match with the *P. furcatus* from Northeast India reported by Roxburgh (1814). The differences are as follows. The Malabar *P. furcatus* grows as a landward species along the streams and rivers, very small size prop roots at the base, stem less branched or unbranched, ellipsoid syncarp with narrow-in-shape, single syncarp per peduncle. The so-reported Northeastern *P. furcatus* grows in open forest with distinct, large prop roots such that it stands on their support, stem well branched with large canopy, syncarp ellipsoid without narrow shape, 5–8 syncarp per peduncle. St. John (1972) reported the species with later description as *P. nepalensis* from Northeastern India. Moreover, St. John (1972) stated that, if a modern collection from Malabar, matching the characters shown in Rheede's plate, could be found, it could be accepted as a neotype of *P. furcatus*. After extensive investigation of the Malabar region of India, we collected a specimen exactly matching the Kaida Tsjeria, described by van Rheede, and confirmed that *P. furcatus* occurs only in the Malabar region of South India and the so-called Northeastern *P. furcatus* is actually *P. nepalensis*.

Specimens examined:

INDIA, Maharashtra, Sindhudurg district, Kudal, alt. 17 m, 16°05′55″ N, 73°42′23″ E, *Rahul Zanan* 01 (pistillate); Maharashtra, Sindhudurg district, Kasal, alt. 54 m, 16°09′08″ N, 73°41′07″ E, *Rahul Zanan 29* (staminate); Goa, South Goa District, Balli, 3–4 km from Balli towards Canacona, alt. 36 m, 15°08′22″ N, 74°01′16″ E, *Rahul Zanan* 44 (pistillate); same locality, *Rahul Zanan* 45 (staminate).

Pandanus unipapillatus Dennst., Schussel 15, 23, 27, 1818; *P. canaranus* Warb., Engler Pflanzenreich IV, 9: 75, fig. 21E, 1900; Martelli, Webbia 4(1): 9, 44, 96, 1913; 4(2): t. 29, figs. 31–35, 1914; C. E. C. Fischer in Gamble, Fl. Pres. Madras 1570. 1931 (3: 1015, 1967, Repr. ed.). Perin Kaida Taddi, Hort Malab. 2: 5, t. 7, 1679; St. John, Bot. Mag. Tokyo 85: 242, f1, 1972; Stone in Sald. & Nicols., Fl. Hassan Distr, 778, 1976; Manilal & Sivar., Fl. Calicut 301, 1982; Manilal & Suresh, New Botanist11: 123, 1984; Sharma et al., Fl. India, 295, 1984. Bhat, Fl. Udupi, 674, 2003.

Original diagnosis:

"Inflorescentia ♂ composite spicata. Spathae inferiores in flagellum spinulosum productae. Spicae dense floribus obtectae, stamina in apice columnae 3 mm longae crassae fasciculata, filamentis minutis, antheris 2 mm longis apice breviter apiculatis. Syncarpium probabiliter ovatum, drupae claviformes fere 5 cm longae 1 cm latae, usque ad apicem connate, apice libero 5–7 mm alto 1 cm lato late convexe pyramidali, stylo 2–2 ½cm lato 3–3 ½mm longo paullo bilobo; mesocarpium medullosum 12 mm longum endocarpium 12 mm longum 8 mm latum."

Expanded diagnosis:

A large shrub to small tree up to 8 m in height, 5–7 cm in diameter; stem more or less erect, smooth, branching near top, few prop roots at base. Leaves attenuate in the apical part and apically on the trunk with unarmed apical ventral plates, up to 1.5–2 m × 3–5 cm, margin with sharp curved prickles and midrib prickly along its whole length. 7–8 spines on midrib (per 10 cm), spines 3 mm long, 15–16 spines on margin (per 10 cm), spines 3 mm long, spines in three rows, yellow, leaf color deep to light green.

Inflorescence terminal, ephemeral, less fragrant, columnar, white to cream colored, 10–12 bracts with variable length, linear lanceolate-lanceolate, margin and midrib with prickles; composite spike, spikes up to 12–20 cm long and dense; 6–8 stamens, androphore 7–9 mm long, stemonophore 3–6 mm long; stamen 5–8 mm long with free up to 1 mm long filament; anther up to 3–5 mm long.

Inflorescence terminal, bracteate, 6–8 with variable length; solitary, oblong-rounded 25–30 cm long, 18–20 cm diameter, carpel simple, compressed, attached to each other, one-seeded drupes, 35–45 mm long, 8–10 mm wide, hexagonal, pileus low rounded, smooth; style 1 cm long, almost at the right angle, bony, brownish; stigma always on the ventral side of the style, 1–2 mm long, forked; endocarp abonic, 9–11 mm long, 5–6 mm diameter; mesocarp fibrous, apical mesocarp 15–17 mm long, basal mesocarp 11–13 mm long; seeds 8–9 mm long, 3–4 mm diameter (Figs. 4.4 and 4.5).

Flowering: July to October, **Fruiting:** July to December.

Leaf micro-morphology: (Fig. 4.6)

In *P. unipapillatus*, the abaxial epidermis is covered by unbranched, randomly distributed papillae. Each papilla is independently borne without stock, 1–3 papillae present per cell, size-wise distinguishable as small triangular to large dome shape (8–10 μm long). The terminal stomatal cells' papillae are branched or less bifurcated and rod shaped (12–20 μm long), partially overarched on guard cells of stomata. Lateral subsidiary cells with 3–6 papillae, forming a wall surrounding stomata. According to Tomlinson (1965) and Kam (1971), abaxial epidermal papillae of *P. unipapillatus* fall under Class 4 (papillose neighboring and subsidiary cells) type of stomata.

Fig. 4.4 *Pandanus unipapillatus*: (**a**) habit; (**b**) staminate inflorescence; (**c**) stamens; (**d**) syncarp; (**e**) stigmas in close view

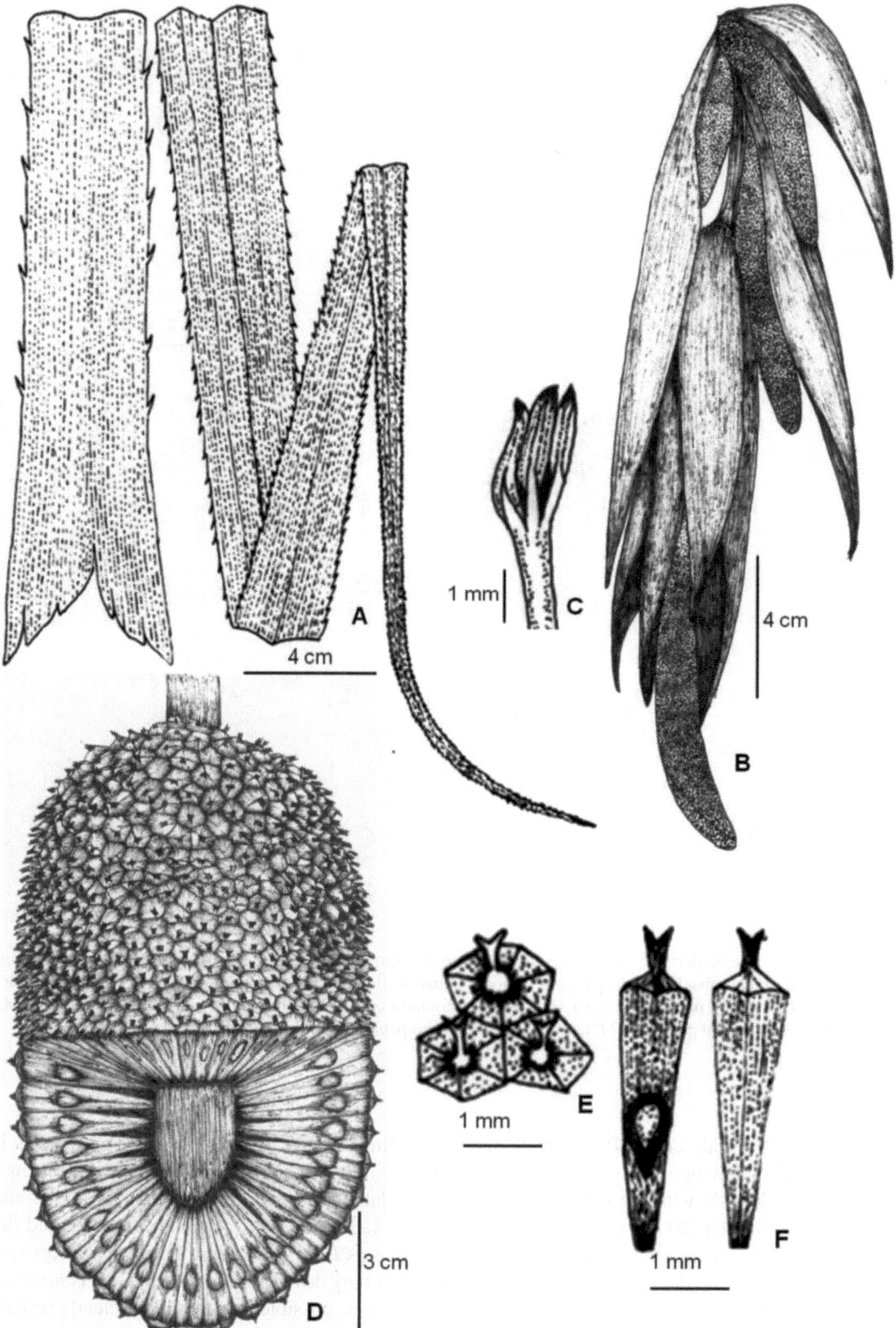

Fig. 4.5 *Pandanus unipapillatus*: (**a**) adaxial and abaxial leaf surface; (**b**) staminate inflorescence; (**c**) androphore; (**d**) syncarp, *lower* half in longitudinal section; (**e**) stigmas in close view; (**f**) single drupes and drupe in longitudinal section showing the mesocarp and endocarp

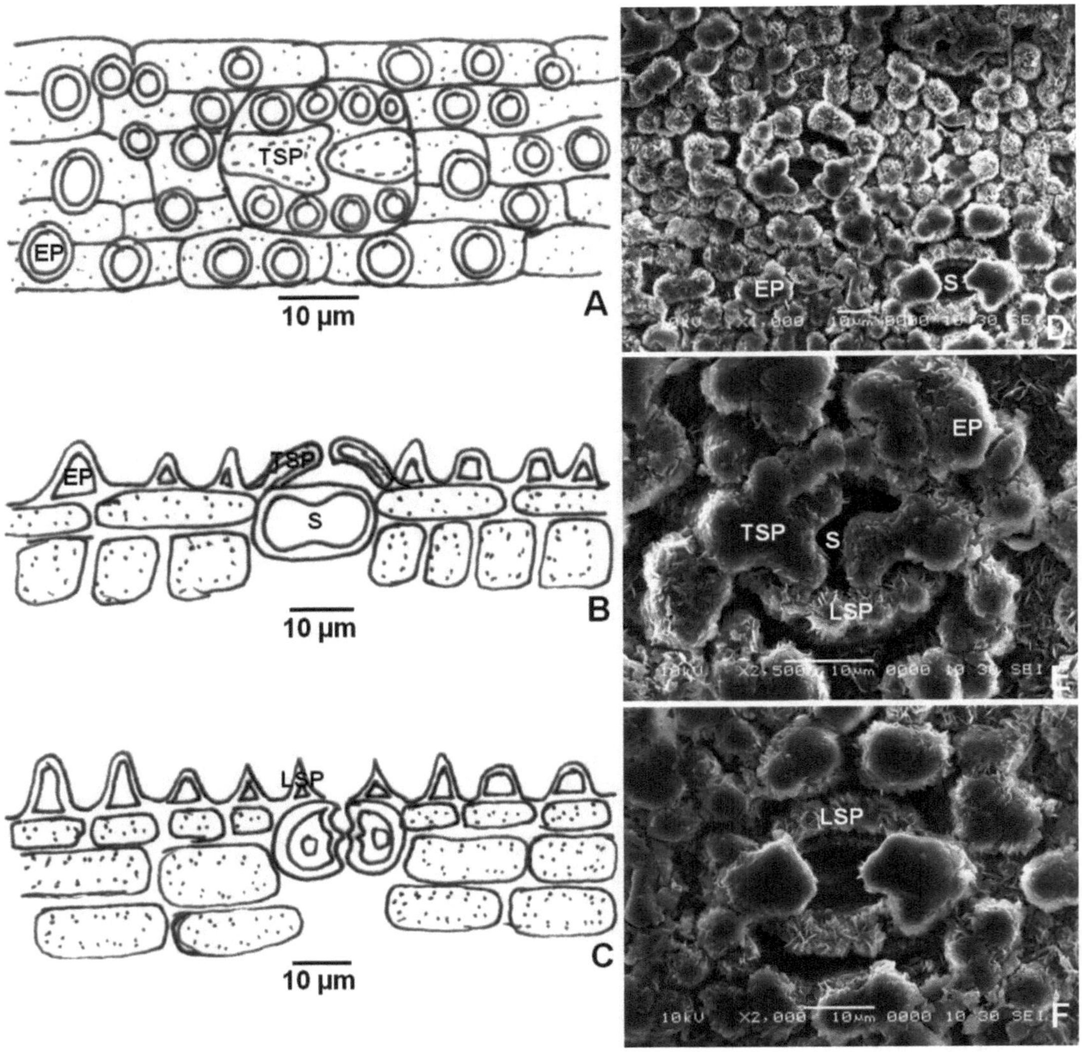

Fig. 4.6 Abaxial leaf epidermal papillae in *P. unipapillatus*: (**a**) surface view; (**b**) transverse section of abaxial epidermis showing terminal subsidiary cells papillae; (**c**) longitudinal section of abaxial epidermis showing lateral subsidiary cell papillae; (**d**) SEM of papillae (×1,000); (**e**) close view around stoma (×2,500); (**f**) close view around stoma (×2,000) (*S* stomata, *EP* epidermal papillae, *TSP* terminal subsidiary cell papillae, *LSP* lateral subsidiary cell papillae)

Notes:

van Rheede (1679) described Perin Kaida Taddi from the Malabar region of India. This species is characterized by having oblong rounded fruit, swollen with lesser nodes separated by deep furrows, surface of each node with a red-blackish papilla with flat surface (van Rheede 1679; Manilal and Suresh 1984; Manilal 2003). van Rheede's Perin Kaida Taddi (1679) is a prelinnean name and the name was only validely published by Bennestedt in 1818 as *P. unipapillatus*. The later species was based on van Rheede's drawing (1679: tab. 7). In 1900, Warburg described a new species, *P. canaranus* Warb. based on Hohenacker 2301 and added *P. unipapillatus* as possible synonym. This clearly reveals *P. canaranus* as a later synonym of *P. unipapillatus*. As suggested by Nicolson et al. (1988) the latter epithet should followed. *Pandanus unipapillatus* is a widely growing endemic species from the Malabar region of South India; common along the stream and river banks throughout Malabar (van Rheede 1679; Manilal and Suresh 1984; Manilal 2003). The species, morphologically, closely resembles *P. furcatus, P. palakkadensis,* and *P. mangalorensis* (Nadaf et al. 2011; Zanan and Nadaf 2012a). All these species are geographically spread in the same region, in open forest with warm humid climatic conditions. Male and female populations of *P. unipapillatus* were observed growing together. Male inflorescence is ephemeral with less fragrance.

Specimens examined:
INDIA: Karnataka, Udupi District, Agumbe Road, alt. 26 m, 13°21′54″ N, 74°52′52″ E, *Rahul Zanan* 02 (pistillate); same locality, *Rahul Zanan* 40 (staminate); Maharashtra, Sindhudurg district, Sawantwadi, 7 km from Sawantwadi towards Londha, alt. 27 m, 15°51′20″ N, 73°50′23″ E, *Rahul Zanan* 7 (pistillate); same locality, *Rahul Zanan* 6 (staminate); Goa Canacona district, Paingen, alt. 7 m, 14°58′02″ N, 74°05′11″ E, *Rahul Zanan* 5 (pistillate); same locality, *Rahul Zanan* 41 (staminate).

Pandanus palakkadensis Nadaf, Zanan & Wakte. Kew Bull. 66: 1–5, 2011.

Original diagnosis:
Pandanus palakkadensis Nadaf, Zanan & Wakte. *P. canarano* Warb.et *P. furcato* Roxb. affinibus sed fructu longo cylindraceo, stylo recto, stigmata oblique acuto, pileo plano, endocarpa humeris elevato.

Expanded diagnosis:
Large dioecious shrub or small tree, about 5–10 m tall, divaricately branched upwards, stem diameter c. 15–20 cm; prop roots absent or rarely present, when present very short at the base of stem; stem erect, rough; leaves linear, c. 300×5–7 cm, both margins and midvein spiny with three rows all along leaf; spines sharp, dense, black in color and pointed; spines at the base 0.4–0.5 cm long, crowded towards apex, ca 0.1 cm long at the tip; marginal spines 1–1.5 cm apart at the base, 0.7–0.1 cm apart at the middle portion and crowded towards the apex; midvein spines 2–2.5 cm apart at the base, 1.5–2 cm apart at the middle.

Staminate inflorescence terminal, compound spike; spathe 100–120 cm long; bracts 9–10, yellowish, linear, lanceolate with variable length, margin and midrib prickled, moderately fragrant; staminate spike 6–15×2 cm, cylindrical; androphore 0.6–0.8 cm long, stemonophore 0.3–0.4 cm long with 8–10 stamens; stamens 0.3–0.4 cm long; filaments 0.1 cm long, arranged in pairs; anther 0.2–0.3 cm long.

Infructescence terminal, 1–2 fruits per infructescence, a solitary syncarp, 45–55×10–12 cm, cylindrical; bracteate, bracts 7–9, linear, lanceolate with variable length, margin and midrib prickled; 1,500–1,600 drupes per infructescence, 3–4×1–1.3 cm, hexagonal, pileus flat, smooth; style 0.75–1 cm long, almost at the right angle to the pileus, bony, brownish; stigma 0.5–0.7 cm long, sharp with oblique apex, brown; endocarp median, abonic, broadly truncate, concave on either side, endocarp with distinctly elevated shoulders, central projection of endocarp attached to stigma, 1–1.2×0.8–0.9 cm, 0.2–0.25 cm thick with 0.7–0.8 cm long tip attached to pileus; mesocarp fibrous, apical mesocarp 0.7–0.8 cm thick, 1.3–1.5 cm thick at base; seed single per locule, ellipsoid, 0.8–0.9×0.3–0.4 cm (Figs. 4.7 and 4.8).

Flowering and fruiting: July to October.

Leaf micro-morphology: (Fig. 4.9)
In *P. palakkadensis,* the whole abaxial epidermis is covered by papillae. The papillae are small, unbranched, randomly distributed, globular (5–8 μm long). The terminal stomatal epidermal cell papillae are bifurcated, elongated, and rod shaped (15–20 μm long), partially overarched on guard cells of stomata. Lateral subsidiary cells with 3–6 papillae, smaller than epidermal papillae, forming a wall surrounding stomata. According to Tomlinson (1965) and Kam (1971), abaxial epidermal papillae of *P. palakkadensis* are categorized under Class 4 (papillose neighboring and subsidiary cells) type of stomata.

Fig. 4.7 *Pandanus palakkadensis*: (**a**) habit; (**b**) staminate inflorescence; (**c**) stamens; (**d**) syncarp; (**e**) stigmas in close view; (**f**) dried syncarp showing drupes

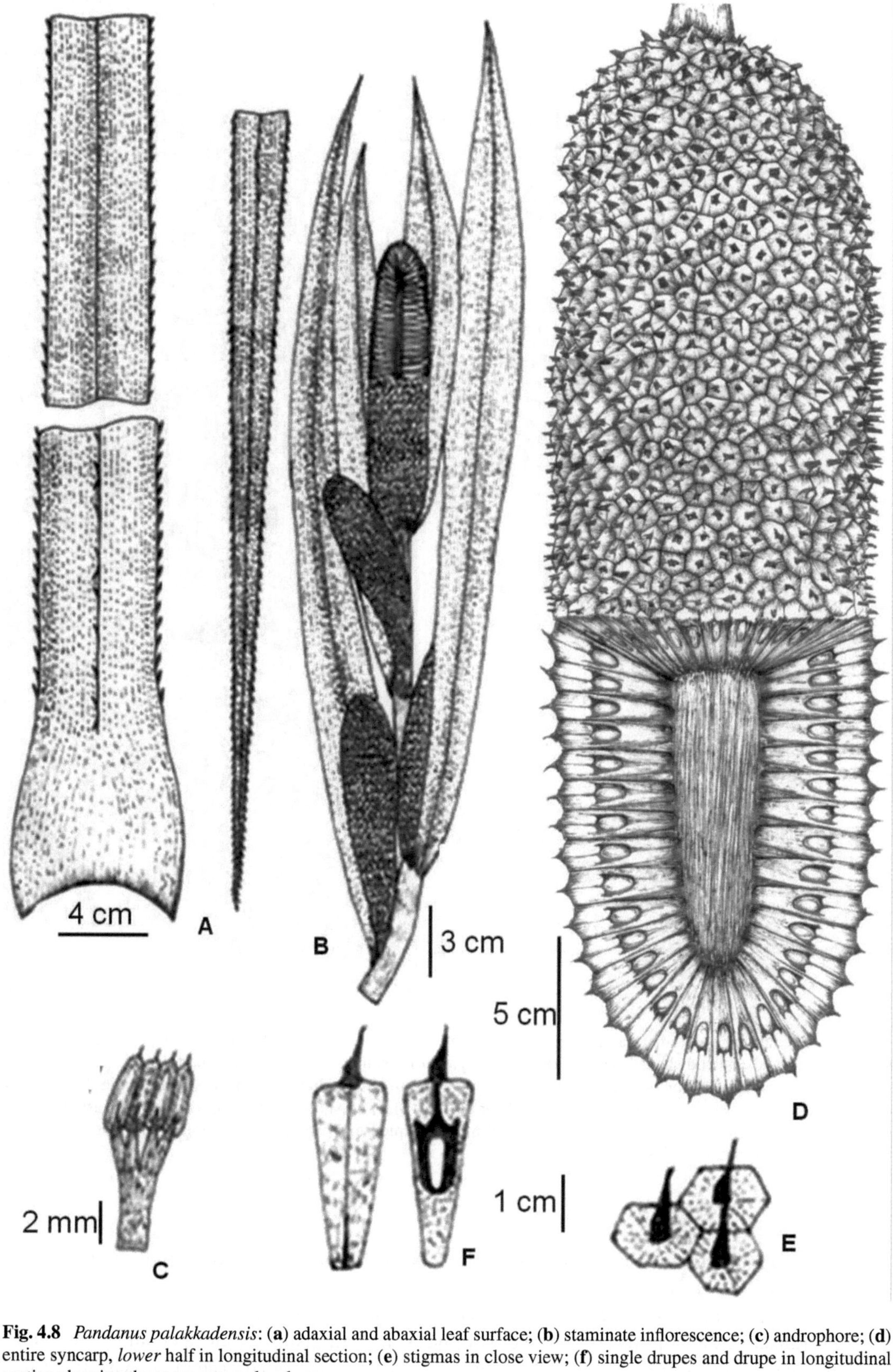

Fig. 4.8 *Pandanus palakkadensis*: (**a**) adaxial and abaxial leaf surface; (**b**) staminate inflorescence; (**c**) androphore; (**d**) entire syncarp, *lower* half in longitudinal section; (**e**) stigmas in close view; (**f**) single drupes and drupe in longitudinal section showing the mesocarp and endocarp

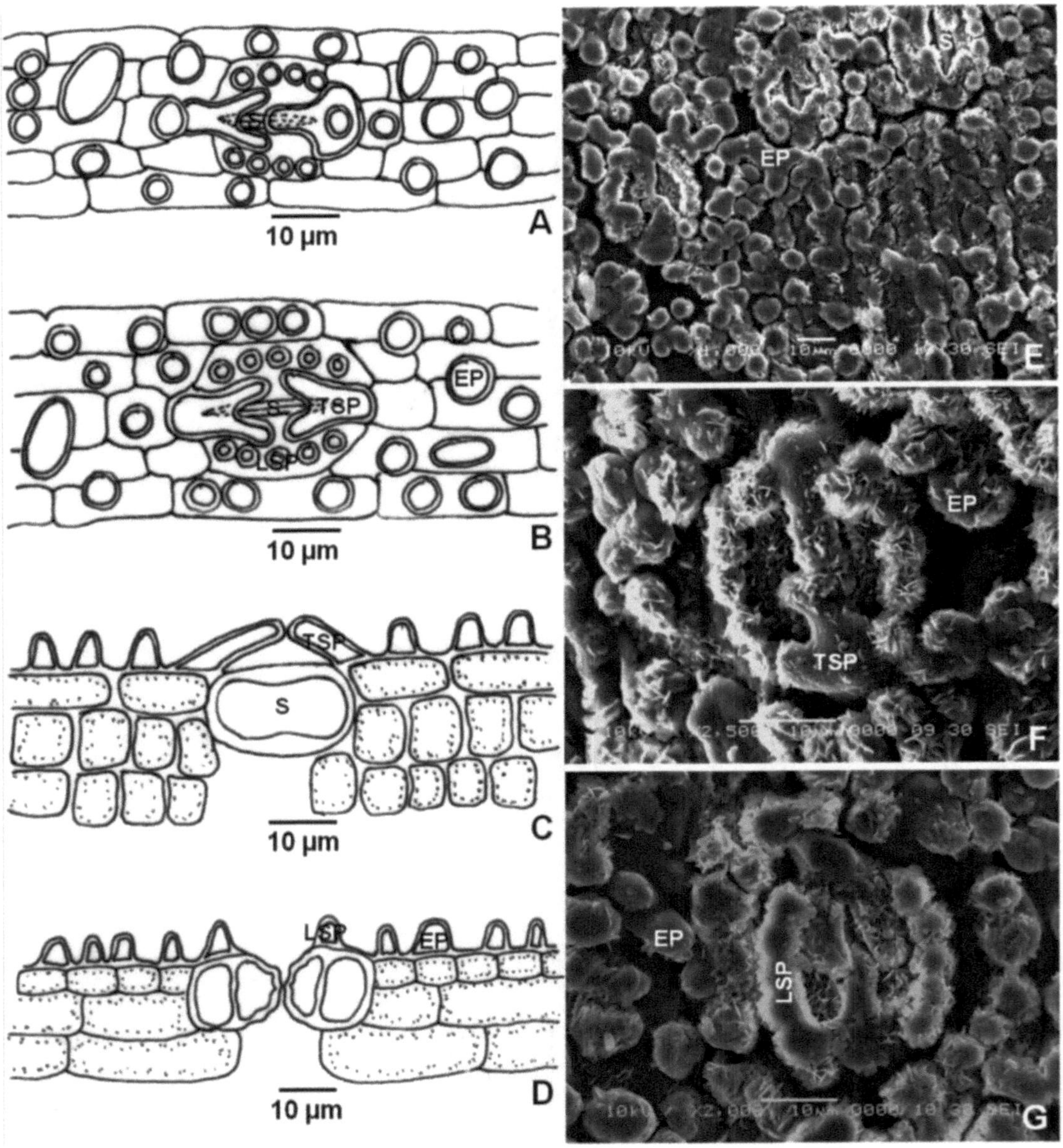

Fig. 4.9 Abaxial leaf epidermal papillae in *P. palakkadensis*: (**a**) surface view of terminal subsidiary cells papillae showing branched and rod-shaped papilla; (**b**) surface view of terminal subsidiary cells papillae showing branched papillae; (**c**) transverse section of abaxial epidermis showing terminal subsidiary cells papillae; (**d**) longitudinal section of abaxial epidermis showing lateral subsidiary cell papillae; (**e**) SEM of papillae (×1,000), (**f**) close view around stoma (×2,500), (**g**) close view around stoma (×2,000) (*S* stomata, *EP* epidermal papillae, *TSP* terminal subsidiary cell papillae, *LSP* lateral subsidiary cell papillae)

Notes:

P. palakkadensis pertains to the subgenus Rykia; it is an endemic species from India and restricted to type locality only. *P. palakkadensis* is distinct from other closely related species, i.e., *P. unipapillatus*, *P. furcatus* and *P. mangalorensis*, in having cylindrical syncarp, drupes with a flat pileus, a broadly truncate endocarp that is concave on either side with distinctly elevated shoulders, and a sharp obliquely pointed stigma. Other *Pandanus* species have ellipsoid or triangular syncarp, drupes with semiorbacular or pyramidal pileus, bilobed or mixed (bilobed and acute) stigma. In India, *P. palakkadensis* is the only species having long, cylindrical syncarp. It is used as a hedge plant along the paddy fields. The male inflorescence is fragrant. At maturity, syncarp turns yellow and then drupes fall down, leaving behind the axis attached to mother plant (Nadaf et al. 2011).

Specimens examined:
INDIA, Kerala, (Palakkad), alt. 40–70 MSL (10°47′57″ N, 67°36′56″ E), *Rahul Zanan* 19 (pistillate) and *Rahul Zanan* 20 (staminate); same locality, *Rahul Zanan* 21 (staminate), *Rahul Zanan* 22 (pistillate).

Pandanus mangalorensis Nadaf & Zanan. Kew Bull. 67: 1–5, 2012.

Original diagnosis:
Pandanus mangalorensis Nadaf & Zanan ramificatio multiplicata promonenti, infructescentia triangulari, drupis osseis triangularibus annulis insidens, apecibus drupis pyramidalibus, stylo acuto bifido, projrctura centrali extensor stylo basi a congeneribus diversa.

Expanded diagnosis:
Medium shrub to large tree, 6–15 m tall, stem erect, rough, 5–8 cm diameter, stem with prominent multiple branching, ending in a clump of scythe-shaped leaves, a few small-size prop roots present at the base, leaves linear, 120–150 cm long, 4–5 cm wide, sheathing at the base, adaxial surface plain and shiny, longitudinal and transverse veins visible on abaxial surfaces; both margins and midrib prickled with three rows all along leaf, prickles brownish to green-yellow colored from base to apex; marginal prickles borne from 4–6 cm above the blade base to the apex, midrib prickled 2–5 cm above the blade base to the apex; marginal prickles at the base 0.3 cm long with 2–4 cm distance and midrib prickles 0.4 cm long with 1–1.5 cm apart; marginal prickles 0.3 cm long with 0.5–0.9 cm apart and midrib prickles 0.3 cm long with 2–3 cm apart at middle portion of leaf blade and crowded towards apex; marginal prickles at the apex 0.1–0.15 cm long with 0.5–0.3 cm apart and midrib prickles 0.1 cm long with 0.3–0.5 cm apart.

Inflorescence terminal, hanging between the clump of leaves; peduncle 60–70 cm long, 1–1.5 cm in diameter, bearing 8–10 yellowish variable length prickled lanceolate bracts, uppermost bracts ~60 cm long, 4–4.5 cm wide in the middle; lowermost 8 cm long, 1.5–2.5 cm wide in the middle; each bract attached with long cylindrical staminate spike, 8–15 cm long, 1.5–2.5 cm diameter, stamens arranged densely in each spike of the inflorescence; staminate spike cylindrical 8–15 × 1.5–2.5 cm, 7–8 stamens, clustered on the 0.3–0.4 cm long and 0.18–0.23 cm wide stemonophore; androphore 0.4–0.6 cm long, 7 anthers united and 1 anther solitary per stemonophore, anthers 0.25–0.3 cm long, oblong, bearing an apical prolongation of the free filaments, 0.1–0.2 cm long, filament base equally expanded to anthers breadth.

Infructescence terminal, hanging between the clump of leaves, single solitary triangular syncarp erect on a straight trigonous peduncle, 16–18 cm long, 11–13 cm diameter, bracteates, bracts linear, lanceolate with variable length, margin and midrib prickled; drupes compressed, 1,100–1,400 drupes per infructescence, drupes arranged on bony 2–4 mm thick triangular ring, triangular ring 5–6 cm wide each side and 3–5 cm long, filled in with spongy tissue, 0.5–1 cm diameter, drupes angular, hexagonal, pileus low pyramidal, smooth, 0.8–0.3 cm high; style acute and bifid, brownish, bony, 0.4–0.6 cm long, almost at right angle to pileus, syncarp with acute bifid styles, each drupe with single style, rarely two drupes fused with separate style, stigma and locule; stigma acute and bifid parallel to style, brown, 0.3–0.5 cm long; endocarp median, abonic, central projection attached to stigma, 1–1.5 long, 0.6–0.9 cm diameter, 0.8–1 cm long tip; apical mesocarp spongy, 1–1.5 cm long; basal mesocarp fibrous, 1–1.7 cm long, 0.4–0.6 cm thick at the base; seed single per locule, oblong, 0.8–1.3 cm long, 0.4–0.7 cm diameter (Figs. 4.10 and 4.11).

Flowering and fruiting: July to October.

Fig. 4.10 *Pandanus mangalorensis*: (**a**) habit; (**b**) staminate inflorescence; (**c**) syncarp; (**d**) transverse section of fruit showing triangular ring and drupes (**e**) stigmas in close view

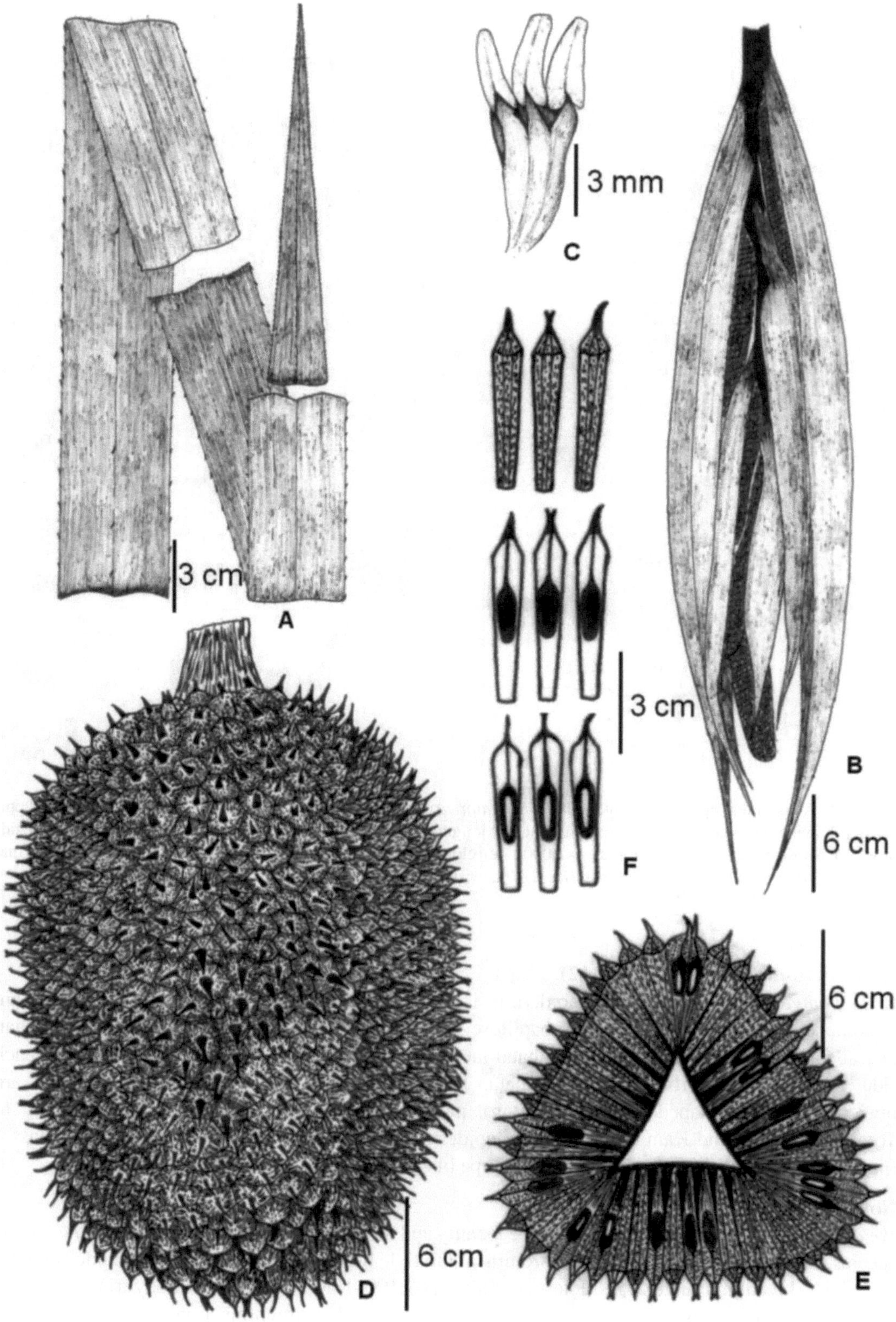

Fig. 4.11 *Pandanus mangalorensis*: (**a**) adaxial and abaxial leaf surface; (**b**) staminate inflorescence; (**c**) androphore; (**d**) syncarp; (**e**) transverse section of fruit showing triangular ring and drupes; (**f**) drupes in longitudinal section showing the mesocarp and endocarp

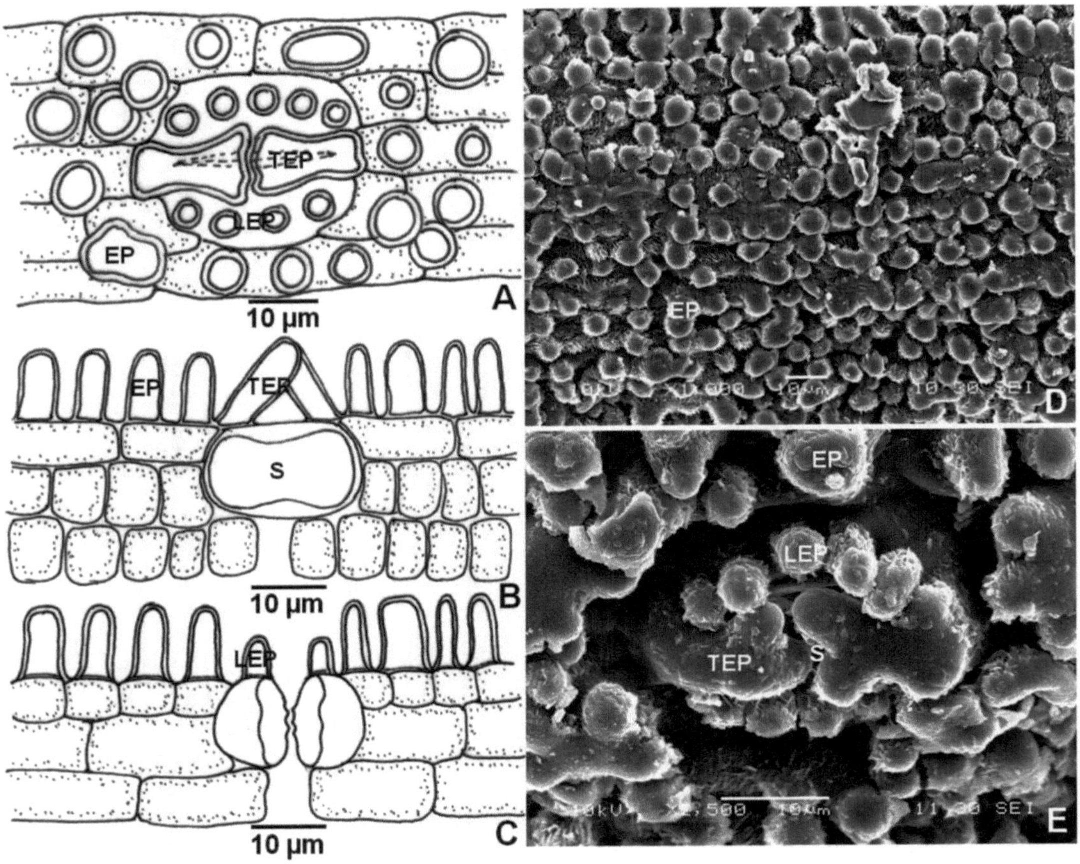

Fig. 4.12 Abaxial leaf epidermal papillae in *P. mangalorensis*: (**a**). surface view; (**b**) transverse section of abaxial epidermis showing terminal subsidiary cells papillae; (**c**) longitudinal section of abaxial epidermis showing lateral subsidiary cell papillae; (**d**) SEM of papillae (×1,000), (**e**) close view around stoma (×2,500) (*S* stomata, *EP* epidermal papillae, *TSP* terminal subsidiary cell papillae, *LSP* lateral subsidiary cell papillae)

Leaf micro-morphology: (Fig. 4.12)

In *P. mangalorensis,* the abaxial epidermis is covered by papillae. The epidermal papillae are unbranched, dome shaped, randomly papillose throughout epidermis, 11–12 µm long, longer than lateral subsidiary cell papillae. The stomatal lateral subsidiary cells are small, 3–6 papillae on each side, forming wall surrounding the stomata. The stomatal terminal subsidiary cells papillae are branched and rod shaped (16–18 µm long), partially overarched on guard cells. According to Tomlinson (1965) and Kam (1971), abaxial epidermal papillae of *P. mangalorensis* showed Class 4 (papillose neighboring and subsidiary cells) type of stomata.

Notes:

Pandanus mangalorensis is endemic to type locality and differs morphologically from other *Pandanus* species in having tall height (6–15 m), prominent multiple branching, filament base expanded equally to anthers' breadth, triangular infructescence, drupes 1,100–1,400, arranged on bony triangular ring, drupe apex pyramidal with acute and bifid style and stigma and central projection extended up to style base. The male inflorescence is ephemeral with average life of 8–12 days after blooming. Based on Stone's infrageneric classification (1974), *P. mangalorensis* can be placed under Pandanus subg. *Rykia* (de Vriese) B.C. Stone, for having leaf apex with unarmed ventral plates; staminate flowers without central pistillode with few stamens, drupes crowed at the apex of stemonophore with free carpel and style acute, bifid (Zanan and Nadaf 2012a).

Specimens examined:
INDIA, Karnataka, (Padil), alt. 6 – 7MSL (12°52′20″ N, 74°53′32″ E) *Rahul Zanan* 30 (pistillate) and *Rahul Zanan* 31 (staminate); same locality, *Rahul Zanan* 32 (staminate), *Rahul Zanan* 33 (pistillate).

Pandanus nepalensis H. St. John, Bot. Mag. Tokyo 85: 241–262, 1972; Karthikyan et al., Fl. India, Florae Indicae Enumeration: Monocotyledonae, 177–178, 1989. *P. furcatus* ver indica Kurz. in J. Bot. 5: 102, 1867; Burkill in Rec., B. S. I. 4: 135, 1910. *P. furcatus* acute non Roxb. Hook. & B. S. I. 6: 484, 1893.

Original diagnosis:
Partes alteres incognitae sunt, infructescentia spicata cum 8 syncarpiis, pedunculo 35 cm longo in apice 25 mm diametro in basi 13 mm diametro clavato trilaterato cernuo cure bracteis deciduis compluribus, rhachide 18 cm longo sinuoso et infra syncarpia bracteato, syncarpio apicali globoso 10 cm diametro, syncarpiis lateralibus 9–9.5 cm longis late ellipsoideis cure 210 drupis, drupis lateralibus et apicalibus 37–42 mm longis 9–12 mm latis 9–10 mm crassis anguste oblongi-lance-oloideis 4–6 – angulosis apice obtuso lateribus inferis planis parte 1/4 supera libera, pileo hemisphaerico paulum 4–6 – anguloso laevi lucido (licet viridi), stylo 5–8 mm longo osseoso brunneo lucido ad laterem obtuto subadscendenti 2–3 mm lato plerumque bifido ¼–1/3 – plo furcis plerumque divergentibus, stylis paucis solum subulatis, endocarpio mediali 26–30 mm longo osseoso brunneo oblanci-ellipsoideo apice cure projectione lanceoloideo ad pileum lateribus lateralibus 1.8–2.5 mm crassis intra laevibus lucidis pallidioribus, semine 16–18 mm longis ellipsoideis, mesocarpio apicali 3–5 mm longo cavernoso cure membranis subbrunneis paucis, mesocarpio basali 6–10 mm longo fibroso et carnoso, drupis basalibus paucis 29–32 mm longis 12–15 mm latis 10–11 mm crassis.

Expanded diagnosis:
Large tree, about 6–8 m height, erect, numerous branched, irregular branching at top only, 25–30 cm diameter, smooth; prop root 4–5 ft long, smooth, 5–8 cm diameter, plant stands on prop root support; leaf long with sheathing at the base, linear, 120–190 cm long and 5–8 cm wide, margins and mid vein spiny with sharp marginal and curved mid vein prickles, spines with three rows on whole leaf, yellow color, on margin at the base 2.5–3 mm long, 5–12 mm distance with10–12 prickles per 10 cm, on midrib at the base 2.7–3 mm long with 0.8–2 cm distance (6–7 prickles per 10 cm), thick, at the middle part 12–13 spines with 3–4 mm long on margin (per 10 cm) with 0.5–2 cm distance, 6–8 spines on midrib with 2–2.5 mm long (per 10 cm) with 1–1.5 cm distance and crowded towards apex.

Inflorescence terminal, compound spike; spathe 200–250 cm long with 12–14 yellowish variable length prickled bracts, low fragrant; staminate spike cylindrical 7–18 cm long, 3–5 cm wide, androphore 1.5–2 cm long, stemonophore 1.2–1.4 cm long with 13–17 stamens; stamens 0.5–0.7 cm long with short filament; anther 0.4–0.6 cm long.

Infructescence terminal with 6–8 solitary globose to oblong-rounded syncarp, with variable length, bracteate with each syncarp, peduncle 35–55 cm long, syncarp 12–14 cm long, 10–12 cm in diameter; drupes compressed, 3.5–4 cm long, 1.5–2 cm wide, hexagonal, pileus conical, pyramidal; style biforked, almost at the right angle of the pileus, bony, brownish, 3–4 mm long; stigma short forked, brown; endocarp median, 1.5–2 cm long, abonic, long tip attached to pileus without shoulders, 1–1.5 mm thick; mesocarp fibrous, apical mesocarp 0.8–1 cm long, basal mesocarp 0.8–1 cm long; Seed single per locule, 1.4–1.9 cm long (Figs. 4.13 and 4.14).

Flowering: June to September **Fruiting:** August to March.

Leaf micro-morphology: (Figs. 4.15 and 4.16)

Fig. 4.13 *Pandanus nepalensis*: (**a**) habit; (**b**) staminate inflorescence; (**c**) stamens; (**d**) bunch of syncarp; (**e**) ripened syncarp; (**f**) stigmas in close view

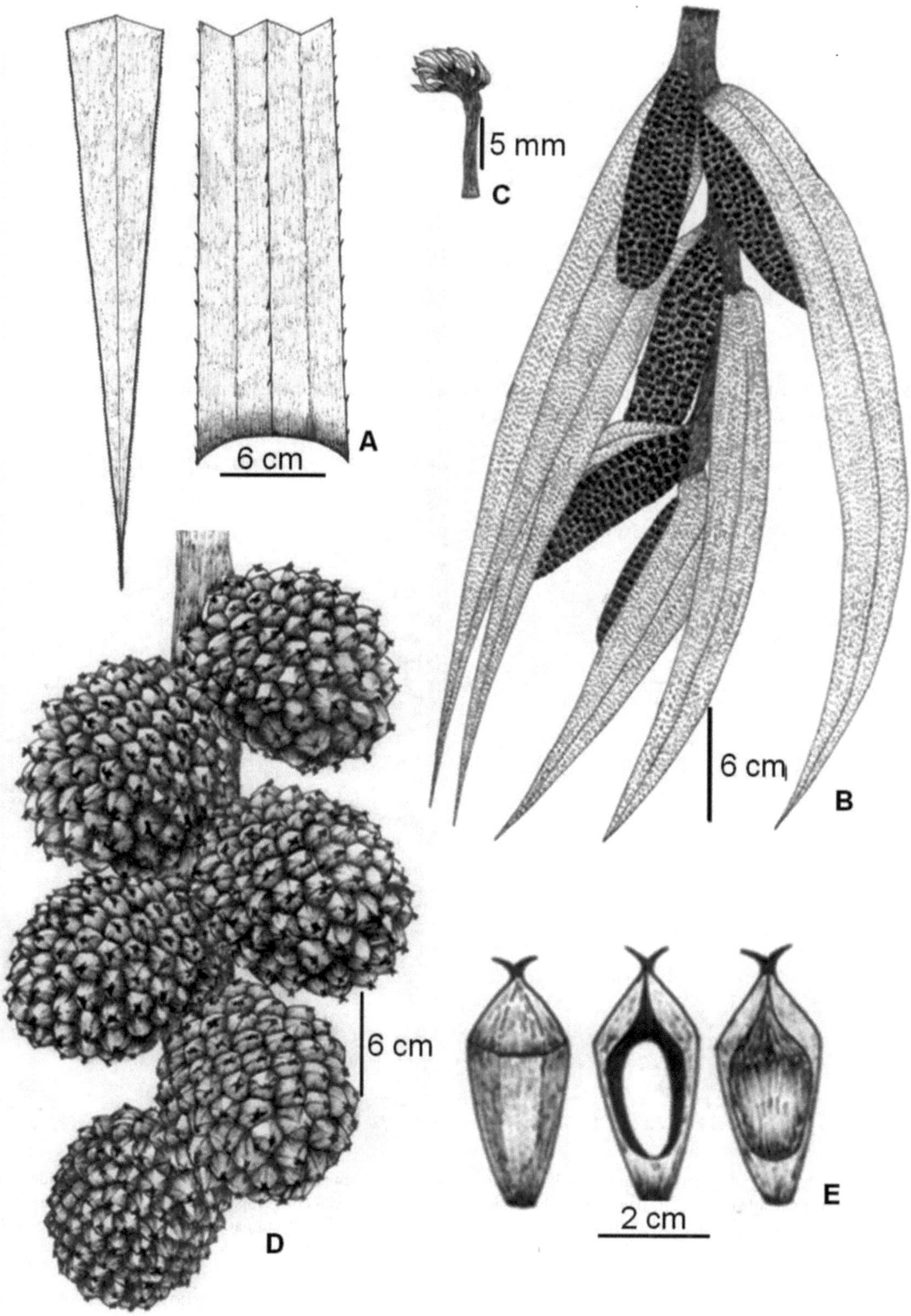

Fig. 4.14 *Pandanus nepalensis*: (**a**) adaxial and abaxial leaf surface; (**b**) staminate inflorescence; (**c**) androphore; (**d**) bunch of syncarps; (**e**) single drupe and drupes in longitudinal section showing the mesocarp and endocarp

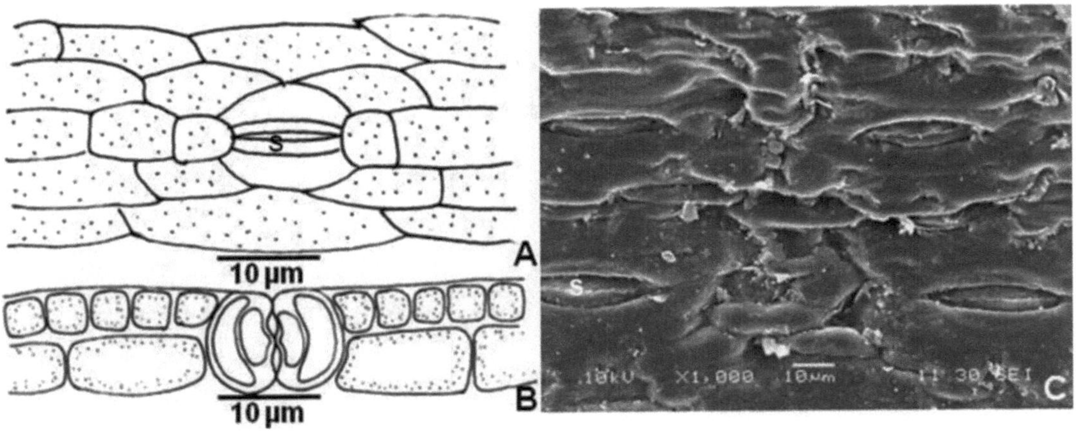

Fig. 4.15 Abaxial leaf epidermal papillae in *P. nepalensis*: (**a**) surface view; (**b**) longitudinal section of abaxial epidermis; (**c**) SEM of abaxial epidermis (×1,000) (*S* stomata)

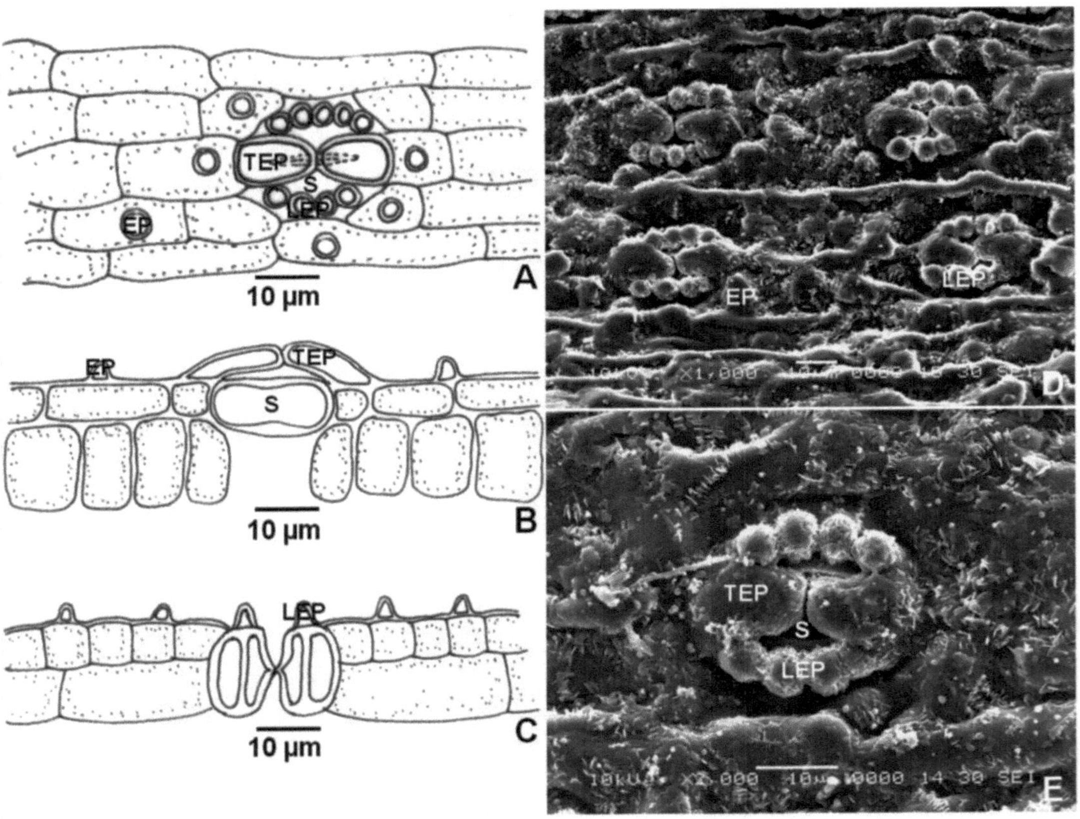

Fig. 4.16 Abaxial leaf epidermal papillae in *P. nepalensis*: (**a**) surface view; (**b**) transverse section of abaxial epidermis showing terminal subsidiary cells papillae; (**c**) longitudinal section of abaxial epidermis showing lateral subsidiary cell papillae; (**d**) SEM of papillae (×1,000); (**e**) close view around stoma (×2,000) (*S* stomata, *EP* epidermal papillae, *TSP* terminal subsidiary cell papillae, *LSP* lateral subsidiary cell papillae)

In *P. nepalensis,* the leaves collected from mature trees in the field did not show the presence any papillae (Fig. 4.15), however, when the young seedlings collected from the field were analyzed, they showed the presence of papillae (Fig. 4.16). They were distributed on subsidiary cells of stomata and rarely on epidermal cells. In this case, terminal subsidiary cells papillae are large, oblong, dome shaped (12–15 μm long), partially overarched on guard cells of stomata. Lateral subsidiary cells papillae are small, simple, and globular or slightly triangular papillae (3–5 papillae on each side). According to Tomlinson (1965) and Kam (1971), abaxial epidermal papillae of *P. nepalensis* fall under Class 1 (unspecialized stomata) and Class 4 (papillose neighboring and subsidiary cells) types of stomata.

Notes:

Pandanus nepalensis has been recorded as one of the dominant species from the North of West Bengal and South districts of Sikkim states of India; popularly known as "Himalayan screw pine." Its closest relative is *P. emarginatus*. Inflorescence of *P. nepalensis* is fragrant, syncarp is pale green in color and turns dark red after maturity.

In eastern Himalayan region only *P. nepalensis* is observed; but has been often wrongly determined as *P. furcatus* by many taxonomists. A detailed discussion is given in the note section of *P. furcatus*. *Pandanus nepalensis* is numerously branched with many long prop roots and multiple syncarps (6–8). *Pandanus emarginatus* from the Northeastern Himalayan region seems to be the close relative of *P. nepalensis*. Both of the species are apparently similar, but can be distinguished from each other through reproductive characters.

Specimens examined:

INDIA, Sikkim, Singtum, alt. 716 m, 27°15′49″ N, 88°33′40″ E, Rahul Zanan 37 (pistillate); same locality, Rahul Zanan 49 (steminate); West Bengal, Darjeeling District, Rambi Bazar, alt. 331 m, 26°58′08″ N, 88°24′33″ E, Rahul Zanan 61 (steminate); same locality, Rahul Zanan 62 (pistillate).

Pandanus emarginatus St. John, Bot. Mag. Tokyo 85: 241–262, 1972; Karthikyan et al., Fl. India, Florae Indicae Enumeration: Monocotyledonae, 177–178, 1989.

Original diagnosis:

Partes alteres incognitae sunt, infructescentia spicata cum 4 syncarpiis, syncarpio apicali 12 cm longo 10 cm diametro latiore ellipsoideo, syncarpiis lateralibus 10–12 cm longis 8–10 cm diametro ovoideo, drupis 45–46 mm longis 13–22 mm latis 13–18 mm crassis cuniformatis 6 – angulosis lateribus inferis planis parte 1/4 supera libera, pileo 5–7 mm alto subconvexo anguloso retuso, stylo 5.5–7 mm longo 2 mm lato bifurcato furcis 2–3 mm longis divergentibus, endocarpio mediali et 27 mm longo osseoso elliptici-oblanceoloideo apice acuto ad stylum extensor lateribus lateralibus 1–1.5 mm crassis, semine 17 mm longo, mesocarpio apicali 10 mm longo, mesocarpio basali 10 mm longo fibroso et carnoso.

Expanded diagnosis:

Large tree, about 6–10 m in height, numerously branched, irregular branching at top, stem erect, smooth, 15–25 cm diameter; prop roots 3–5 ft long from base to top, smooth, 5–8 cm diameter, whole plant stands on prop root support; leaf long with sheathing at the base, linear, 150–200 cm long and 6–7 cm wide, margins and mid vein spiny with sharp prickles, margins and midrib spiny with three rows of whole leaf, margin with sharp prickles and midrib with curved prickles at the base, prickles yellow, on margin at the base 3–4 mm long, 5–12 mm distance with10–12 prickles per 10 cm, on midrib at the base 2–3 mm long with 1–4 cm distance (2–4 prickles per 10 cm), thick, at the middle part 2–3 mm long on margin (14–17 spines per 10 cm) with 4–10 mm distance, on midrib 2–2.5 mm long (2–4 spines per 10 cm) with 2–4 cm distance and crowded towards apex.

Inflorescence terminal, compound spike; spathe 180–220 cm long with 11–12 yellowish variable length prickled bracts; staminate spike cylindrical 6–15 × 2–4 cm, androphore 1–1.3 cm long, stemonophore 0.6–0.8 cm long with 12–16 stamens with short filament; anther 0.4–0.5 cm long.

Infructescence terminal with 4–5 solitary oblong-rounded syncarp, variable length, 13–16 cm long, 10–14 cm in diameter, bracteates, peduncle 35–55 cm long; drupes compressed, 3.5–4.5 cm long, 1.5–2.2 cm wide, 6 angled, upper ¼ th free, pileus low pyramidal or semiorbicular and retuse at the center; style 4–5.5 mm long, 2 mm wide, biforked, brownish bony; stigma short forked, brown, 2–3 mm long; endocarp abonic, median, bony, 2.5–3 cm long, 1–1.5 mm thick, central projection attached to pileus without shoulders; mesocarp fibrous, apical mesocarp 0.8–1 cm long, basal mesocarp 0.9–1.2 cm long; Seed single per locule, 2.3–2.8 cm long (Figs. 4.17 and 4.18).

Flowering: June to September, **Fruiting:** August to March.

Leaf micro-morphology: (Fig. 4.19)
In *P. emarginatus*, the abaxial epidermis is covered by papillae. They are randomly distributed, separately borne, 1–2 papillae per cell, size-wise distinguishable as small (6–7 μm long, 5.5–6.5 μm width) and large dome-shaped papillae (11–12.2 μm long and 8–11 μm width). The terminal subsidiary cells papillae are elongated, unbranched, overarched on guard cells of stomata (11–15 μm long). The lateral subsidiary cells papillae are 3–5, small unbranched, low triangular in nature (3.8–5 μm long). According to Tomlinson (1965) and Kam (1971), *P. emarginatus* showed Class 4 (papillose neighboring and subsidiary cells) type of stomata.

Notes:
Pandanus emarginatus is a member of the section *Rykia*, endemic to type locality inflorescence is moderately fragrant, syncarp is pale green in color and turns blackish red at maturity. It is closely related to *P. nepalensis*. *P. nepalensis* has linear leaf, 120–190 cm long, 5–8 cm wide; 12–14 prickled bracts with 13–17 stamens per stemonophore; 6–8 solitary globose to oblong-rounded syncarp, 12–14 cm long, 10–12 cm in diameter, drupes 3.5–4 cm long, 1.5–2 cm wide with pyramidal pileus, style bifid, 3–4 mm long, seed 1.4–1.9 cm long. *P. emarginatus* has linear leaf, 150–200 cm long, 6–7 cm wide. 11–12 prickled bracts with 12–16 stamens per stemonophore; 4–5 solitary oblong-rounded syncarp, 13–16 cm long, 10–14 cm in diameter, drupes 3.5–4.5 cm long, 1.5–2.2 cm wide with semiorbicular pileus and retuse at the center, style bifid 4–5.5 mm long, seed 2.3–2.8 cm long.

Specimens examined:
INDIA, Arunachal Pradesh, West Kameng District, Bhalukpung, alt. 170 m, 27°00′39″ N, 92°38′37″ E, *Rahul Zanan* 38 (pistilate); same locality *Rahul Zanan* 46 (staminate).

Pandanus diversus H. St. John, Bot. Mag. Tokyo 85: 241–262, 1972; Karthikyan et al., Fl. India, Florae Indicae Enumeration: Monocotyledonae, 177–178, 1989.

Original diagnosis:
Arboriformis 5 in alta est, (radicibus et caulibus incognitis), foliis licet 1.7–2 m longis proxima basem 5.2 cm latis proxima medium 6 cm latis coriaceis supra viridibus lucidis infra pallide viridibus U-sulcatis 2-plicatis in sectione mediali cum 60 nervis parallelis secundariis in dimidio quoque nervis tertiis transversis omnino conspicuis et reticulis oblongis formantibus lamina late ligulata (apice interito) basi inermi pallida, ex fete 15 cm marginibus cum aculeis 2 mm longis 7–18 mm separatis subulatis salientibus stramineis, midnervo infra ultra 30 cm inermi, proxima medium marginibus cum aculeis 3–8 mm separatis biformatis majoribus 2.5–3 mm longis arcuatis crasse subulatis crebre adscendentibus et cure illis minoribus 1–1.5 mm longis arcuatis subulatis adscendentibus alternantibus, midnervo vicinali infra inermi, infructescentia cum syncarpio solitario, pedunculo 25 vel plus cm longo 2 cm diametro trilaterato cum bracteis foliosis lateralibus compluribus, bracteis apicalibus syncarpium inclusis, syncarpio fere 14 cm longo 9 cm diametro ellipsoideo trilaterato, drupis 30–34 mm longis 5–7 mm latis 3–6 mm crassis fusiformibus 5–6 angulosis parte 1/2 supera libera, pileo 10–12 mm alto ovoideo ad lanceoloideo 5–6 anguloso sublaevi, stylo 6–8 mm longo lucido proxime curvato plerumque bifido cum dentibus 1.5–2.5 mm longis plerumque salientibus, drupis

Fig. 4.17 *Pandanus emarginatus*: (**a**) habit; (**b**) bunch of syncarps (**c**) single syncarp; (**d**) stigmas in close view

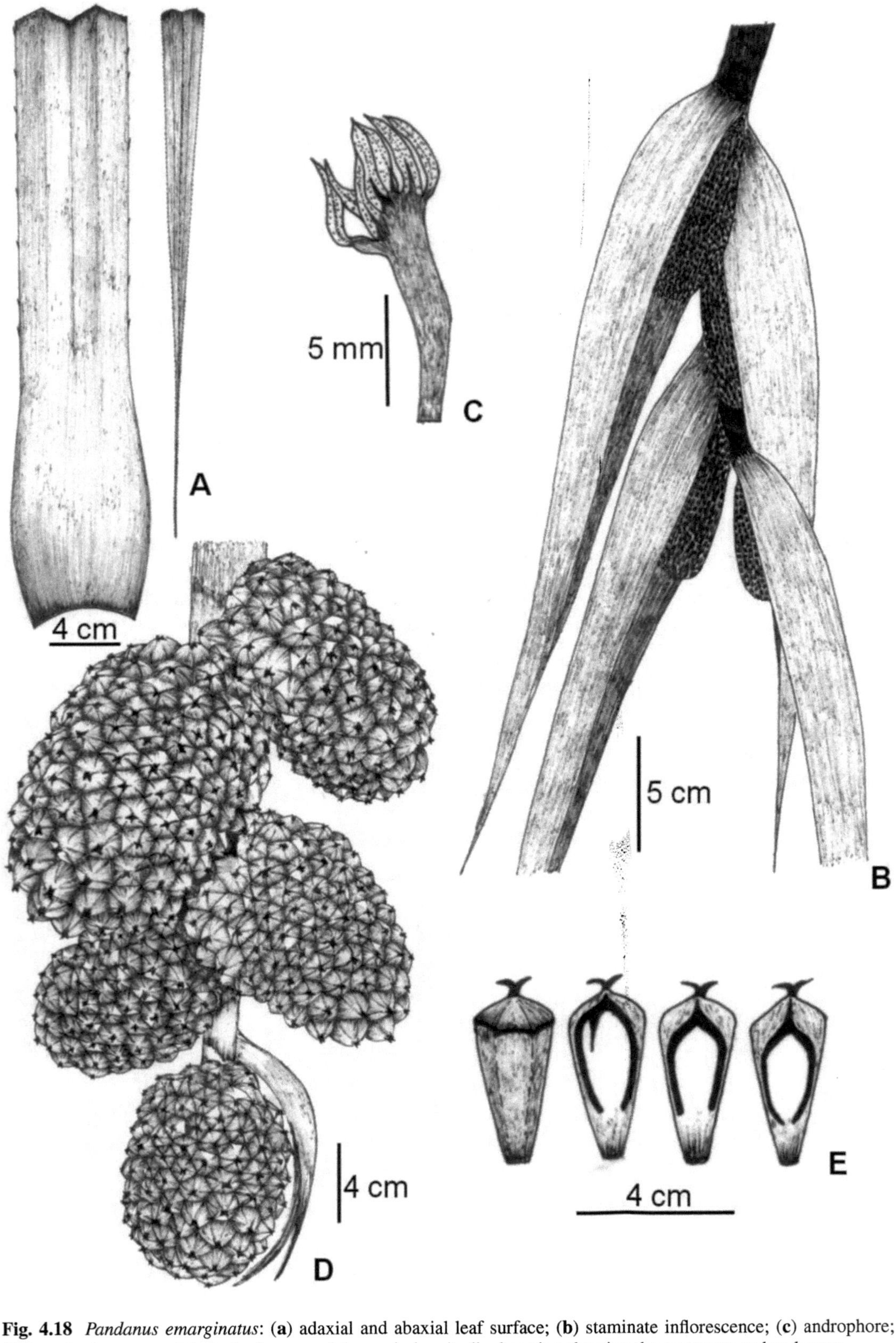

Fig. 4.18 *Pandanus emarginatus*: (**a**) adaxial and abaxial leaf surface; (**b**) staminate inflorescence; (**c**) androphore; (**d**) bunch of syncarps; (**e**) single drupes and drupes in longitudinal section showing the mesocarp and endocarp

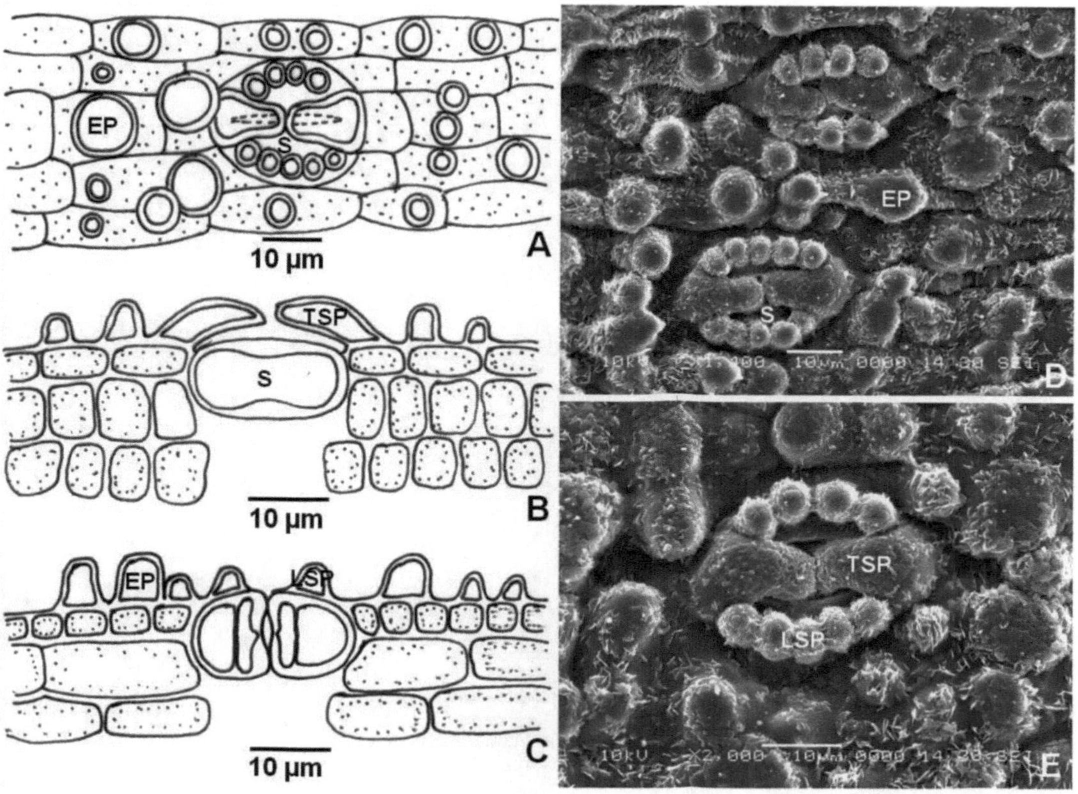

Fig. 4.19 Abaxial leaf epidermal papillae in *P. emarginatus*: (**a**) surface view; (**b**) transverse section of abaxial epidermis showing terminal subsidiary cells papillae; (**c**) longitudinal section of abaxial epidermis showing lateral subsidiary cell papillae; (**d**) SEM of papillae (×1,000); (**e**) close view around stoma (×2,000) (*S* stomata, *EP* epidermal papillae, *TSP* terminal subsidiary cell papillae, *LSP* lateral subsidiary cell papillae)

paucis cum stylo arcuato crasse subulato, stigmate 3.5–4 mm longo lanci-lineari integro vel bifido pallid brunneo papilloso, endocarpio inframediali 20–22 mm longo lineari-lanceoloideo acuminato osseoso extra pallido intra ferruginoso lateribus lateralibus 0.5 mm crassis intra rugosis lucidis, semine 6 mm longo ellipsoideo, mesocarpio apicali 9–10 mm longo cum parenchyma ct fibris, mesocarpio basali 2 mm longo fibroso et carnoso.

Expanded diagnosis:

Large shrub to small tree, about 3–6 m in height, unbranched, small suckers at the base, stem erect, rough, 10–15 cm diameter; prop roots at the base only, 2–3 ft long from base to top, rough, 4–5 cm diameter, plant stands on prop root support. Leaf long, sheathing at the base, linear, 180–250 cm long and 5–6 cm wide, margins and mid vein spiny with sharp spines, adaxial epidermis green, shiny, abaxial epidermis pale green, basal part 6–8 cm of leaf brown, margins and midrib spiny with three rows of whole leaf, margin with sharp spines and midrib with curved spines at the base, white, spines start from 10–15 cm leaf base, spines on basal margin 3–4 mm long, 5–10 mm distance, on midrib 2–4 mm long, 7–20 mm distance, thick; marginal spines at the middle part 1–2 mm long with 3–5 cm distance (20–25 spines per 10 cm), midrib spines 2–3 mm long with 1–3 cm distance (5–7 spines per 10 cm) and crowded towards apex.

Inflorescence: Not recorded. Infructescence: Terminal, solitary syncarp, bracteates with variable length (40 cm to 180 cm), white, peduncle 25–40 cm long, 2 cm in diameter, 3 sided, syncarp

Fig. 4.20 *Pandanus diversus*: (**a**) habit; (**b**) syncarp; (**c**) stigmas in close view showing acute stigma; (**d**) stigmas in close view showing bifid stigma

ellipsoid, pale green in color and turns red at maturity, 27–33 cm long, 18–22 cm in diameter; drupes compressed, 6–8 cm long, 1.6–2 cm wide, hexagonal or angular, 5–6 angled, pileus pyramidal, upper 2–2.5 cm high, free, smooth; style linear with 8–13 mm long, bony, shining, curved, majority bifid with 1.3–2 mm long, some acute with 1.3–1.7 cm long, brown; stigma 4–7 mm long, linear, entire or bifid, pale brown; endocarp median, bony, linear, with acute apex reaching the style without shoulders, 3.5–4 cm long, 1–2 mm thick; mesocarp fibrous, apical mesocarp 1.4–1.6 cm long, basal mesocarp 2–2.2 cm long; seed 3.4–3.8 cm long, single per locule (Figs. 4.20 and 4.21).

Flowering: Not recorded **Fruiting:** October to March.

Leaf micro-morphology: (Fig. 4.22)
In *P. diversus* the abaxial epidermis is covered by papillae. The papillae are unambiguously distributed, unbranched, separately borne, each papilla acquired single cell breadth with 1–3 dome-shaped papillae per cell, attached each other forming vertical row, opposite to cell length having

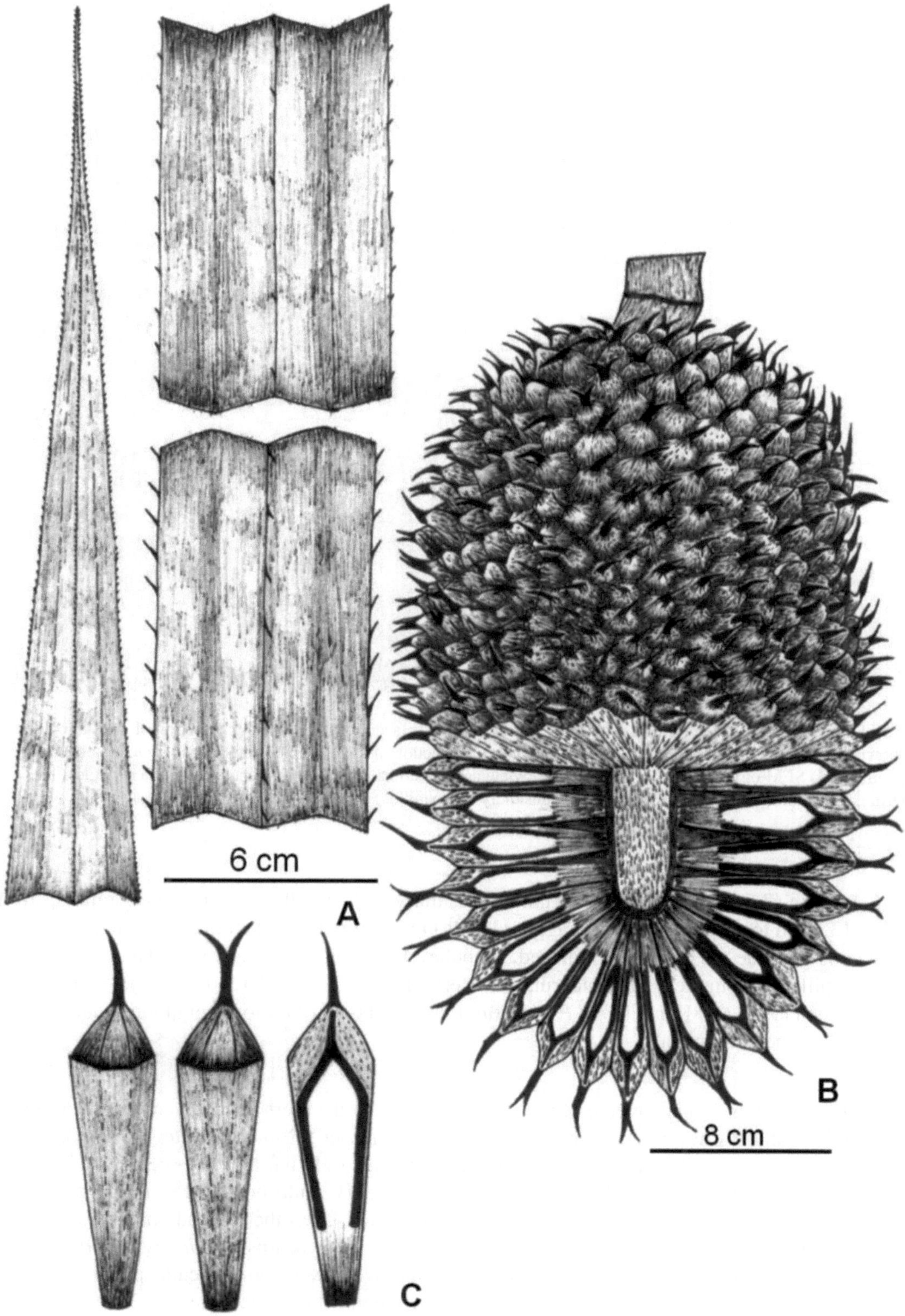

Fig. 4.21 *Pandanus diversus*: (**a**) adaxial and abaxial leaf surface; (**b**) syncarp, *lower* half in longitudinal section; (**c**) drupes with details of stigma and drupe in longitudinal section showing the mesocarp and endocarp

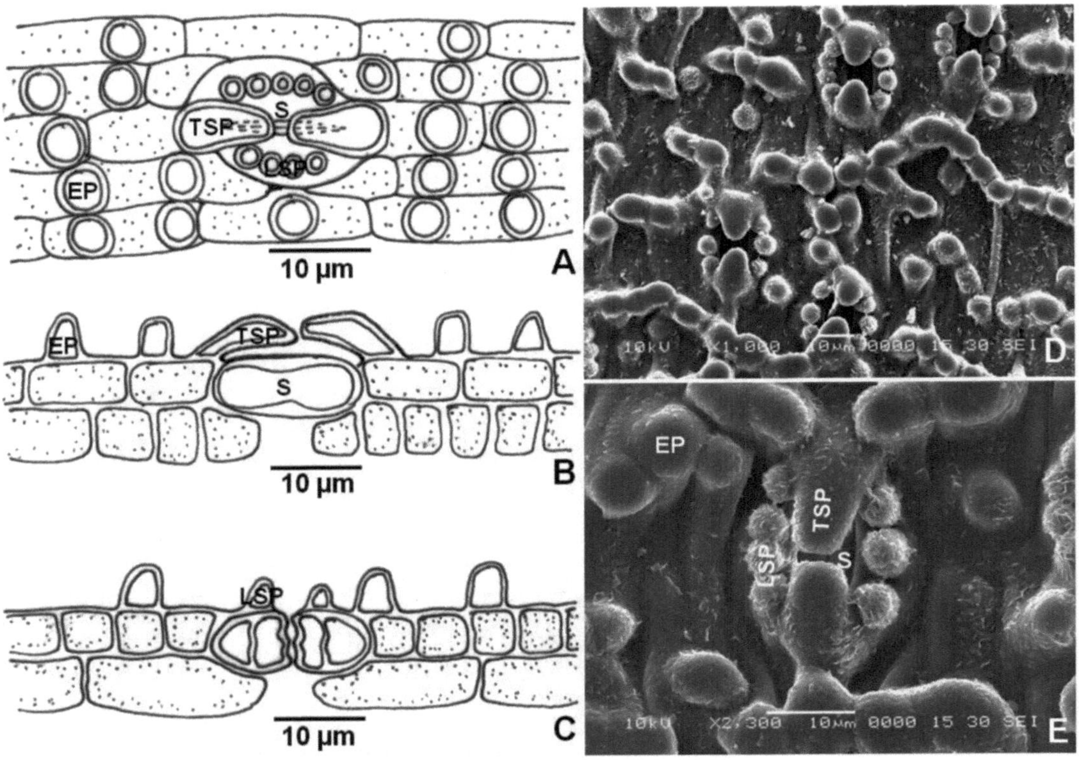

Fig. 4.22 Abaxial leaf epidermal papillae in *P. diversus*: (**a**) surface view; (**b**) transverse section of abaxial epidermis showing terminal subsidiary cells papillae; (**c**) longitudinal section of abaxial epidermis showing lateral subsidiary cell papillae; (**d**) SEM of papillae (×1,000); (**e**) close view around stoma (×2,300) (*S* stomata, *EP* epidermal papillae, *TSP* terminal subsidiary cell papillae, *LSP* lateral subsidiary cell papillae)

3–6 papillae (7.5–10 μ long). The terminal subsidiary cells papillae are elongated, ranging from 12.5–13.5 μm long, partially covering guard cells (11–14 μm long) of stomata. The lateral subsidiary cells papillae are small, unbranched, globular, 4–5 papillae on each side; lateral subsidiary cells papillae are smaller than epidermal and terminal papillae (9–10.5 μm long). According to Tomlinson (1965) and Kam (1971) *P. diversus* showed Class 4 (papillose neighboring and subsidiary cells) type of stomata.

Notes:

Pandanus diversus is endemic to terrestrial region of Northeastern Himalayan region of India. Morphologically *P. diversus* shows resemblance with *P. furcatus, P. unipapillatus, P. palakkadensis,* and *P. mangalorensis*, which are distributed in Malabar region of India. All these species are the members section Rykia. *P. diversus* haslong prominent prop roots, multi-branched stem, large syncarp, pileus pyramidal with two types of styles (bifid and acute). The other related species have either small-size prop roots or absent, less branched stem with small syncarps. *Pandanus furcatus* showed pyramidal pileus with bifid style, *Pandanus unipapillatus* showed semiorbicular pileus with bifid style, *P. palakkadensis* showed flat pileus with obliquely pointed or acute style and *P. mangalorensis* showed elevated or semiorbicular pileus with bifid and acute style.

Specimens examined:
INDIA, Assam, Silchar District, Dalu, alt. 35 m, 24°55′06″ N, 92°51′01″ E, *Rahul Zanan* 36 (pistillate); Assam, Silchar District, Koyla Godam Village, alt. 48 m, 24°24′00″ N, 92°51′12″ E, *Rahul Zanan* 42 (pistillate).

Pandanus unguifer Hook. f., Bot. Mag. 104: t. 6,347, 1878; St. John., Bot. Mag. Tokyo 85: 241–262, 1972; Karthikyan et al., Fl. India, Florae Indicae Enumeration: Monocotyledonae, 177–178, 1989. Pandanus minor Buch. Ham. in Wall. Cat. nr 8,592, 1832, nom. nud; Solms in Linnaea 52: 18, 1879; Hook f. Fl. Brit India 6: 485, 1893.

Original diagnosis:
Humilis, caule gracile prostrato, foliis subdistichis 2–3 pedalibus 11/2-2 poll. latis lineari-loratis junioribus abrupte senioribus sensim caudato-acuminatis marginibus et costa subtus spinoso-dentatis, syncarpio sessili suberecto ovoideo diametro pugilli, drupis obovoideo-cuneatis monospermis lateribus angulatis vertice hemispherico laevissimo medio ungue parvo duro nitido acuto v. emarginato v. 2 - dentato v. bicorni abrupte terminatis.

Expanded diagnosis:
Small shrub with simple, prostrate, decumbent stems 2–4 ft long, about 2.5 cm in diameter; prop roots very small at the base or on prostrate stem; leaves linear, sub-distichous, ensiform, 100–200 cm long, 4–5 cm wide at the base, 5–6 cm wide at the middle and apical 1/10 abruptly narrowed, bearing a 3–5 cm long subulate tip, margins and mid vein spiny with sharp spines, yellow, basal spines thick and apical thin, 10–15 cm marginal base unarmed, on margin at the base 2–3 mm long, 2–3.5 mm distance with 4–5 spines (per 10 cm), on midrib 2–4 mm long with 1.5–4 cm distance, at the middle part 7–8 spines with 1–1.5 mm long on margin (per 10 cm) with 1.5–2 cm distance, 2–3 spines on midrib with 2–3 mm long (per 10 cm) with 4–5 cm distance and crowded towards apex.

Inflorescence terminal, compound spike; spathe 6–22 cm long with 6–8 yellowish variable length spiny bracts, moderately fragrant; staminate spike 6–10 cm long, androphore 0.6–0.7 cm long, stemonophore 0.4–0.6 cm long with 8–10 stamens; stamens 0.4–0.5 cm long, filaments 0.2–0.3 cm; anther 0.3–0.4 cm long.

Terminal infructescence with globular syncarp, 7–10 cm long peduncle; prickled bracts; drupes, 3–3.5 cm long, 1.5–2 cm wide, hexagonal or 5–6 angled, pileus conical, 0.7–1 cm high, smooth; style curve ascending, spreading lobes or merely subulate and unlobed, 2–3 mm long; stigma 1.5–2.5 mm long, bony, brown; endocarp concave on either side, endocarp with distinctly elevated shoulders, apex with a conic projection to the style, 1.5–1.8 cm long, 0.9–1.3 cm wide, walls 1.5–2.5 mm thick; mesocarp fibrous, apical mesocarp 0.8–1 cm long, basal 4–5 cm long; seed single per locule, 1–1.2 long, 0.8–1 cm wide (Figs. 4.23 and 4.24).

Flowering: July to September, **Fruiting:** August to March.

Leaf micro-morphology: (Fig. 4.25)
In *P. unguifer* abaxial epidermal papillae are absent. According to Tomlinson (1965) and Kam (1971) *P. unguifer* showed Class 1 (unspecialized stomata) type of stomata.

Notes:
J. D. Hooker (1878) for the first time described *P. unguifer* from the Mungpoo hills of the Darjeeling district of West Bengal. Later on it was considered as *P. minor* Buch. Ham. by Hooker (1893) and subsequently followed as *P. minor* by most taxonomists (Srivastava 1997). Warburg (1900) and Martelli (1913) rejected this species and reduced it as the synonym of *P. unguifer*. Both reported species were based on the same Hooker's material from Khasia (Noltile 1994). St John (1972) reported

Fig. 4.23 *Pandanus unguifer*: (**a**) habit; (**b**) staminate inflorescence; (**c**) stamens; (**d**) syncarp; (**e**) stigmas in close view

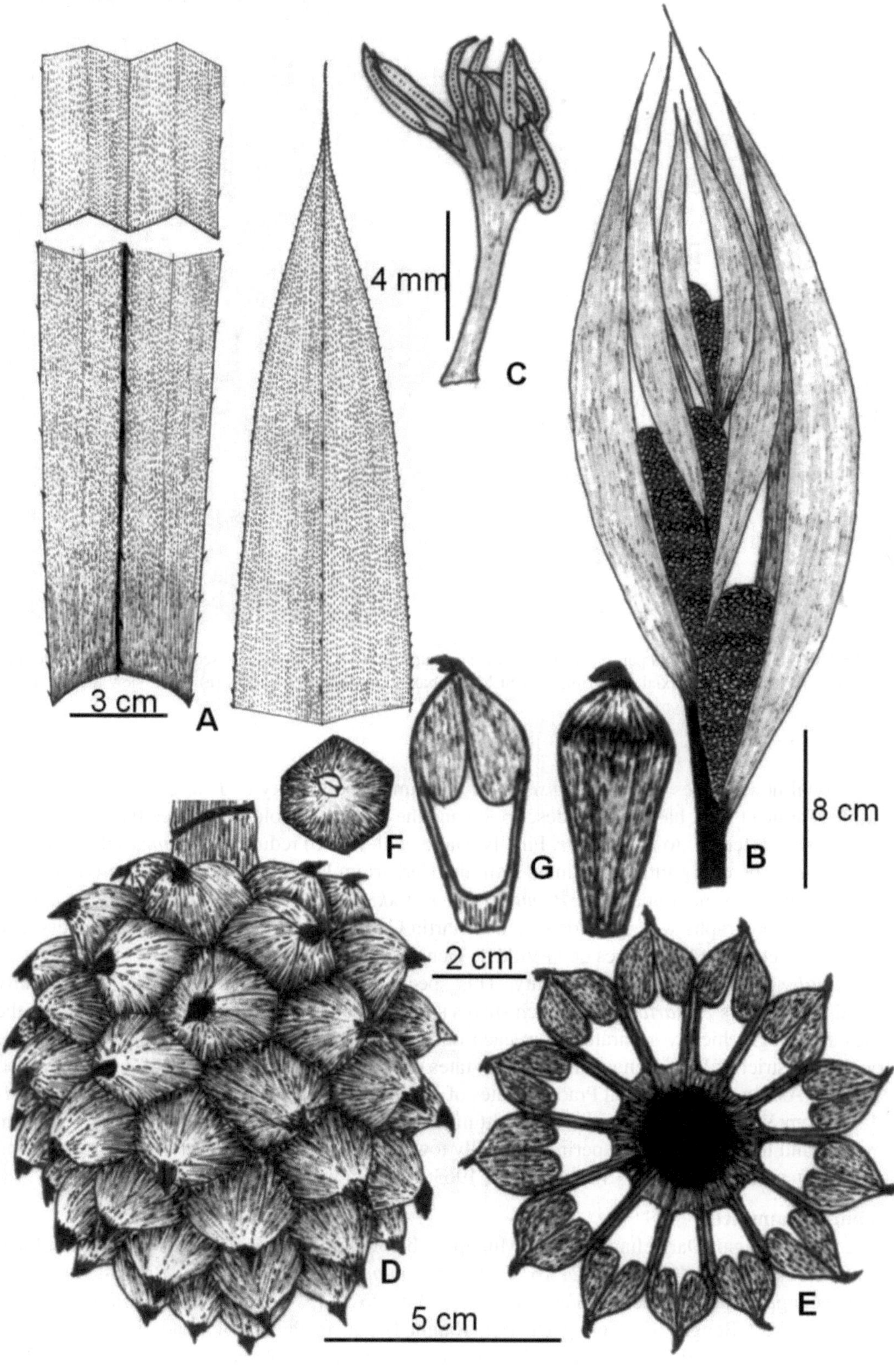

Fig. 4.24 *Pandanus unguifer*: (**a**) adaxial and abaxial leaf surface; (**b**) staminate inflorescence; (**c**) androphore; (**d**) syncarp; (**e**) transverse section of syncarp; (**f**) stigmas in close view; (**g**) single drupe and drupe in longitudinal section showing the mesocarp and endocarp

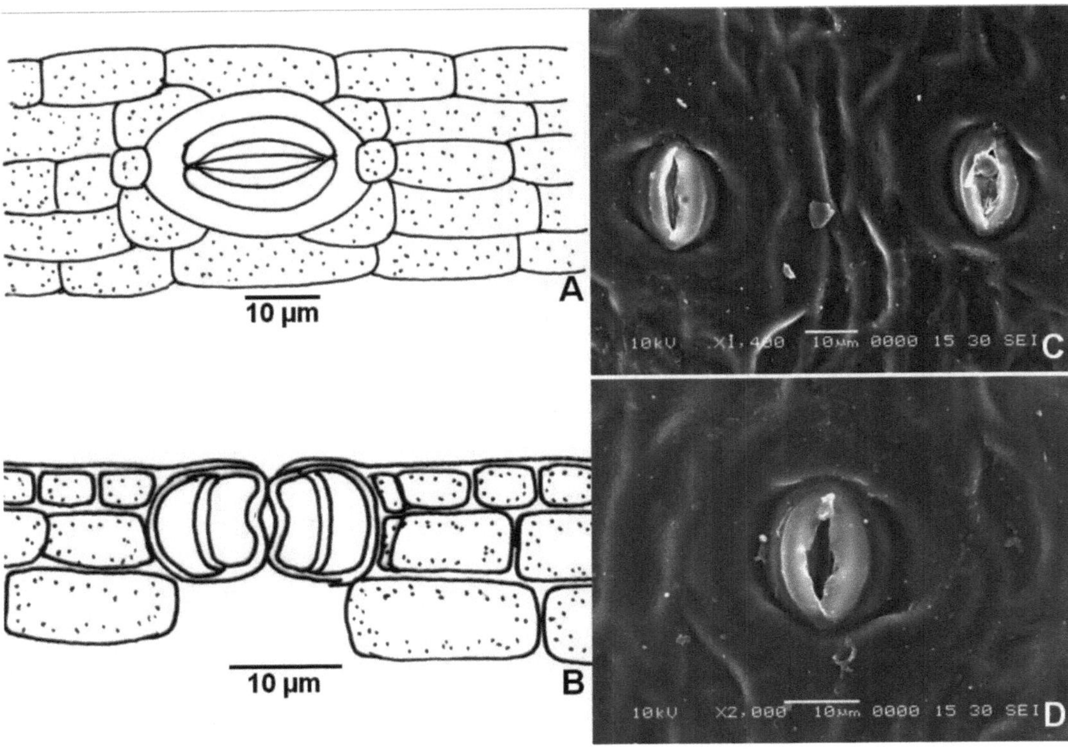

Fig. 4.25 Abaxial leaf epidermal papillae in *P. unguifer*: (**a**) surface view; (**b**) transverse section of abaxial epidermis; (**c**) longitudinal section of abaxial epidermis; (**d**) SEM of abaxial epidermis (×1,400); (**e**) close view around stoma (×2,000) (*S* stomata)

an additional new species - *P. sikkimensis* from the same type locality of *P. unguifer* along with *P. unguifer*. Noltile (1994), based on the description and the locality of holotype, stated that *P. sikkimensis* seems to be referable to *P. unguifer*. Finally, Hajra et al. (1996) reduced *P. minor* as the synonym of *P. unguifer*. The type fruiting specimen from the Herbarium Universitatis Florentinae (Herbarium No. 2,698, Finland) mentioned that *P. minor* and *P. sikkimensis* are the synonyms of *P. unguifer* (Herbarium photographs were provided by Dr. Martin Callmander, Missouri Botanical Garden and Conservatoire et Jardin botaniques de la Ville de Genève).

P. unguifer is endemic to type locality. This species morphologically resembled our recently reported new species *P. martinianus*, which shares some leaf and infructescence characters. Both the species are geographically separated from each other; *P. unguifer* is found at high elevation in the Darjeeling District of West Bengal and Sikkim states of India and *P. martinianus* is found in lowland areas in the Assam and Arunachal Pradesh states of India. It also closely resembles *P. multidrupaceus* St. John from Vietnam, *P. longicaudatus* Holttum & St. John from Malaya, and *P. bifidus* St. John from Thailand in having leaves tapering gradually towards the tip and ending in a subulate apex and some infructescence characters (St. John 1962, 1963; Holttum and St. John 1962).

Specimens examined:

INDIA. West Bengal, Darjeeling District, Mungpoo, below Sureil Bungalow in Chinchona cultivation, alt. 1,514 m, 26°58′09″ N, 88°21′16″ E, *Rahul Zanan* 35 (pistillate); West Bengal, Darjeeling District, Mungpoo, near Chinchona Factory, alt. 1,143 m, 26°58′13″ N, 88°22′20″ E, *Rahul Zanan* 23 (steminate); West Bengal, Darjeeling District, Rambi Bazar, in the forest, road side, alt. 293 m, 26°59′17″ N, 88°25′09″ E, *Rahul Zanan* 24 (pistillate); same locality, *Rahul Zanan* 25 (steminate).

Pandanus martinianus Nadaf & Zanan, Phytotaxa 73: 2.

Original diagnosis:
Pandanus martinianus is distinguished from other Indian *Pandanus* species in having 10–12 cm long, 6–7 cm wide ellipsoid syncarp, compiling 140–160 drupes, upper 2/3 bifid style and lower 1/3 unlobed style, endocarp basal, cone shaped, thin wall, 0.1–0.15 cm thick, inner side of endocarp is shiny with flat apex.

Expanded diagnosis:
Small shrub, 1–3 ft tall with slender decumbent stem, 1.5–2.5 cm in diameter, smooth, branched from the base; prop roots present, small, at the base only. Leaves coriaceous, on the distal part of branches, linear, 80–120 cm long, 4–5 cm wide near the sheath, 3–4 cm wide in the middle, abruptly attenuate in its distal 1/10, ending by a subulate tip 3–4 cm long, auricles lacking, adaxial leaf surface dark green and abaxial surface pale green, longitudinal and transverse veins visible on both surfaces; leaf margins and abaxial mid vein spiny with sharp prickles, yellow, proximal prickles thick becoming thinner in the distal end; marginal prickles appear at 5–7 cm from the base of leaf and extending towards apex, prickles on margin at the base of leaves 1–3 mm long, 0.4–1.5 cm apart (8–10 spines per 10 cm), on midrib 2–2.5 mm long and 1–1.5 cm apart, at the middle part of the leaf prickles 1–2 mm, 0.3–1 cm apart (10–12 spines per 10 cm), on midrib 1–1.5 mm long 1–3 cm apart (3–4 spines per 10 cm) and crowded towards apex.

Infructescence terminal, a solitary syncarp 10–12 cm long, 6–7 cm wide, ellipsoid; peduncle 10–16 cm long, 1–1.5 cm wide, straight, trigonous, veins visible, bearing 7–8 bracts, prickled, pale yellow; 140–160 drupes per syncarp, connate in the syncarp, drupes 2.7–3.2 cm long, 1.5–2 cm wide, hexagonal, pileus conical, superior portion free, 0.7–1 cm high, V-shaped, smooth; style curved; bifid and unlobed, 2–3 mm long, upper 2/3 portion of syncarp with bifid styles and lower 1/3 portion with unlobed styles; single stigma per drupe in the center of the conical pileus, 1.5–3 mm long, bony, brown; endocarp cone shaped, inner side smooth and shiny, apex flat with a conic projection extending to the style base, 0.8–1 cm long in the center of drupe, 0.4–0.6 cm wide at middle part, 0.1–0.15 cm thick; mesocarp fibrous, superior mesocarp narrow and compact, 0.3–0.5 cm long, inferior mesocarp thick and fibrous, 1.2–1.4 cm long; single seed per locule, seed locule conical, 0.6–0.8 cm long. Staminate flower unknown (Figs. 4.26 and 4.27).

Fruiting: June to September.

Leaf micro-morphology: (Fig. 4.28)
In *P. martinianus,* abaxial epidermal papillae are absent. According to Tomlinson (1965) and Kam (1971), *P. unguifer* showed Class 1 (unspecialized stomata) type of stomata.

Etymology:
The species is named in honor of Dr. Martin W. Callmander, assistant curator and technical advisor for the Conservation and Research Program in Antananarivo, Madagascar (Missouri Botanical Garden and Conservatoire et Jardin botaniques de la Ville de Genève) for his contributions to the knowledge of the family Pandanaceae, especially in Madagascar and New Caledonia and for his tireless support in the current revision of the Indian screw pines.

Note:
Pandanus martinianus closely resembles *P. unguifer* in having a single small syncarp with small leaves abruptly attenuated in the distal part. It can be distinguished from *P. unguifer* in having ellipsoid syncarp (vs. globular in *P. unguifer*), 6–7 × 10–12 cm (vs. 7–10 × 7–10 cm), with 140–160 drupes (vs. 90–110 drupes), with upper 2/3 bifid style and lower 1/3 unlobed style (vs. unlobed bifid style). Furthermore, both species are geographically separated from each other. *P. martinianus* is found in Assam and Arunachal Pradesh in open forest, whereas *P. unguifer* is endemic to West Bengal and Sikkim states of India in dense forests. *Pandanus martinianus* is placed in *Pandanus* subg. *Rykia* (de

Fig. 4.26 *Pandanus martinianus*: (**a**) habit; (**b**) staminate inflorescence; (**c**) stamens; (**d**) syncarp; (**e**) stigmas in close view

Vriese) B.C. Stone following the infrageneric classification of Stone (1974), for having leaf apex with unarmed ventral plates; drupes crowned with free carpel and acute or bifid style and stigmas on the ventral side of style (Zanan and Nadaf 2012b).

Pandanus martinianus is also compared with Asian *Pandanus* species and we observed that it closely resembles *P. multidrupaceus* St. John from Vietnam, *P. longicaudatus* Holttum & St. John from Malaya, and *P. bifidus* St. John from Thailand, having leaves tapering gradually towards the tip and ending in a subulate apex and some infructescence characters. *P. multidrupaceus* has narrowly obovoid-cylindric syncarp, 22 cm long, 12 cm diameter with 752 drupes with irregularly bifid style, endocarp in lower 1/3, thick, ellipsoid. *P. longicaudatus* has elliptic-ovoid syncarp 16–17 cm long, 10.5 cm diameter with about 900 drupes, apical style entire and remaining bifurcate, endocarp in lower 1/3, thick, ellipsoid. *Pandanus bifidus* has ellipsoid syncarp, 12 cm long, and 10 cm in diameter bearing very numerous drupes, endocarp median, obovoid (St. John 1962, 1963; Holttum & St. John 1962).

Specimens examined:

India, Arunachal Pradesh, West Siang District, Foot hill of Along, alt. 294 m, 28°10′16″ N, 94°46′37″ E, 29th September 2010, *Rahul Zanan* 17 (pistillate); same locality 29th September 2010, *Rahul Zanan* 65 (pistillate); Assam, Dhimaji District, Dhimaji, alt. 107 m, 27°28′22″ N, 94°32′31″ E, 29th September 2010, *Rahul Zanan* 18 (pistillate); same locality, 29th September 2010, *Rahul Zanan* 34 (pistillate); same locality, 29th September 2010, *Rahul Zanan* 66 (pistillate); same locality, 29th September 2010, *Rahul Zanan* 67 (pistillate).

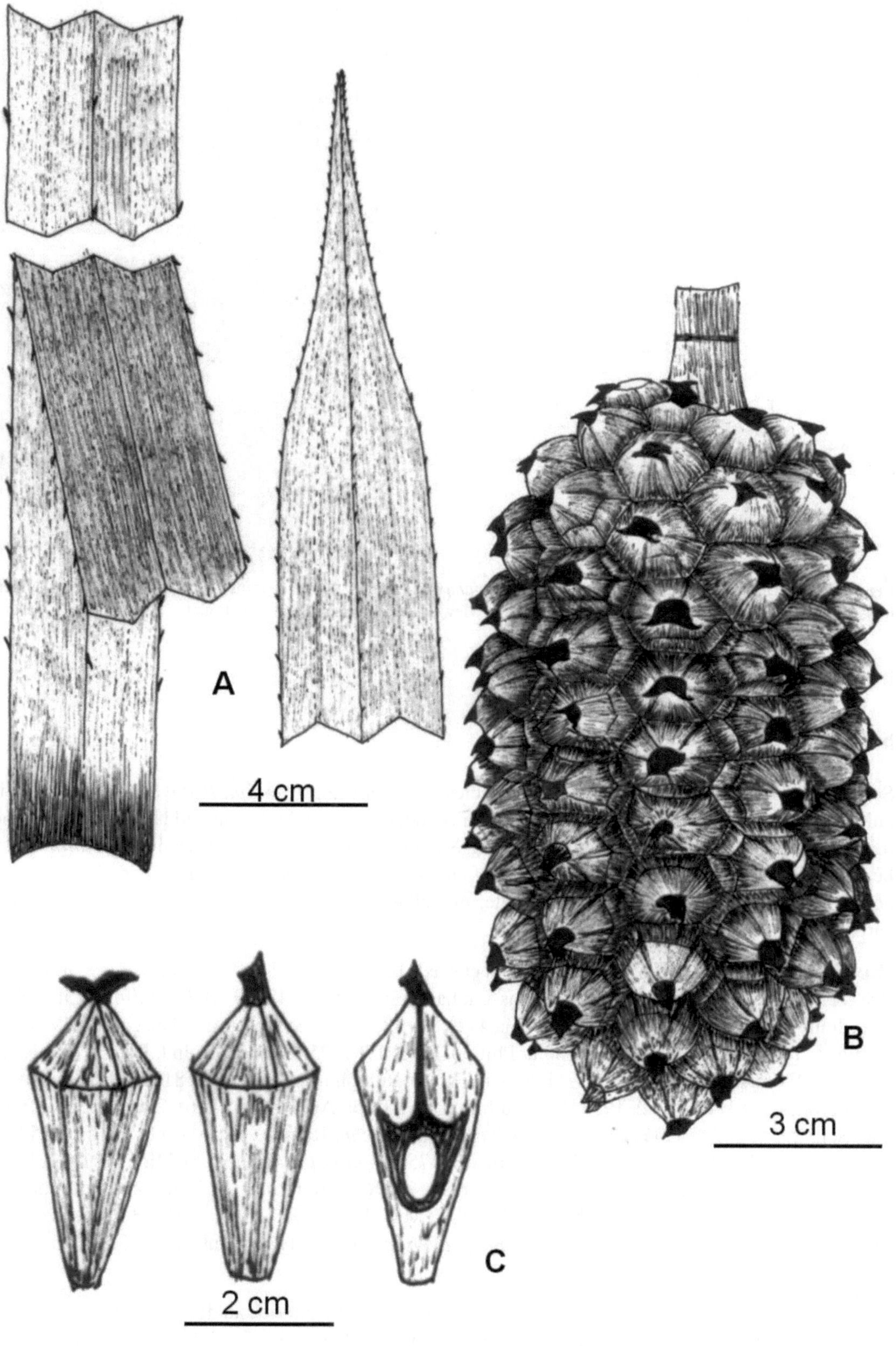

Fig. 4.27 *Pandanus martinianus*: (**a**) adaxial and abaxial leaf surface; (**b**) syncarp; (**c**) drupe in longitudinal section showing the mesocarp and endocarp

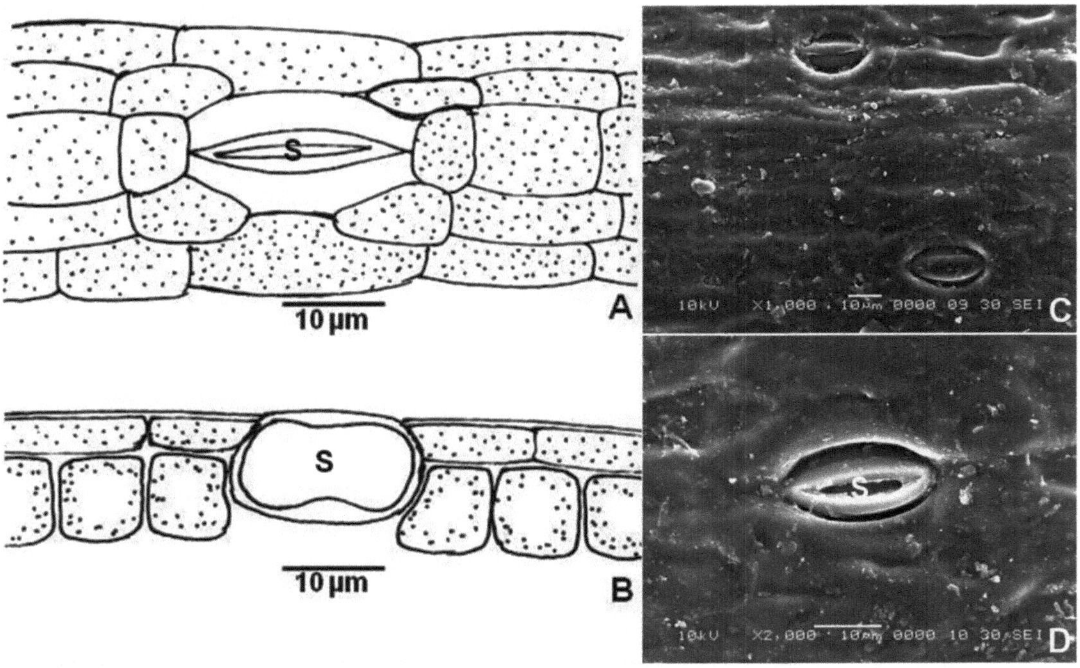

Fig. 4.28 Abaxial leaf epidermal papillae in *P. martinianus*: (**a**) surface view; (**b**) transverse section of abaxial epidermis; (**c**) SEM of abaxial epidermis (×1,000); (**d**) close view around stoma (×2,000) (*S* stomata)

Section *Hombronia* (Gaudich.) Warb.

Section *Hombronia* is represented by two species, *P. dubius* and *P. leram*. *P. dubius* is exotic and is cultivated in gardens of Southern India. *P. leram* is endemic to the Andaman and Nicobar Islands and it is cultivated at the botanical garden, Botanical Survey of India, Yarcud. The section *Hombronia* can be recognized by the following combination of characters: leaf epidermis clearly divided into costal-costal and landward zones; stamens connate, umbellate to racemose; style ovate-acute or deltoid; carpel free or connate in phalanges.

Pandanus dubius Spreng., Syst. Veg. of Linnaeus, ed. 16 by Spreng. 3: 897, 1826; S. Kurz, Jour. Bot. Brit. & For. 5: 127–128, t. 64, 1867; Solms-Laubach, Linnaea 42: 48–49, 1878; Balfour, Linn. Soc. Bot., Jour. 17: 45–46, 1878; Martelli, Webbia 4 (1): 12, 48, 94, 1913; Warburg, Pflanzenreich IV, 9: 50, 52, fig. 1, 14 A – D, 1900; Merrill, Enum. Philipp. PI. 1: 17, 1925; Miquel, FI. Ind. Bat. 3: 159, 1855; Koorders, Exkursfl. Java 1: 79–80, 1911; Merrill, Interp. Rumphius's Herb. Amb. 8182, 1917; Kanehira, FI. Micronesica 61–62, fig. 2, 1933; Kyushu Imp. Univ., Dept. Agric., Jour. 4 (6): 260, 1935; St. John, Pac. Sci., 29: 4, p. 371–406, 1975. *Folium Baggea maritimum* Rumph., Herb. Amb. 4: 151, 1750. *P. latifolius* Rumph., Herb. Amb. 4: 146–147, pI. 78, 1750. *P. latifolius* Rumph. ex Miq., FI. Ind. Bat. 3: 164, 1855, non Perrotet (1825). *P. edulis* De Vriese, in Hook. Kew Jour. 6: 264, 1854, non Thouars (1808). *P. latissimus* Blume, Rumphia 1: t. 53, 1835; Miq., PI. Jungh. 1: 166, 1851–53. *P. Bagea* Miq., FI. Ind. Bat. 3: 159, 1860, based on *Folium Baggeaverum* Rumph., Herb. Amb. 4: 150, 1750.

Original diagnosis:

P. ramis fimplicibus, fpinis foliorum deciduis, fructu globofo folitario, nucibus obconicis distinctis. *Amboina. Inf. Mafcaren.* (Folium baggeae maritimum Rumph. 4. t. 80. 75. ?. P. erigens Thuars.?)

Expanded diagnosis:
Tree, up to 15 m in height, 15–30 cm in diameter; more or less erect stem, smooth, pale brown, branching from base to top, prop roots at the base, 2–5 m long, about 7–10 cm in diameter. Leaves linear with unarmed apical ventral plates, up to 2.5–3 m long, 15–20 cm wide at the middle, apical tip abruptly narrowed and bearing a subulate 7–11 cm long tip, adaxial surface glossy, coriaceus, above dark green and shiny, below pale green and less shiny, two pleated, margin with sharp curved prickles, midrib prickly along its whole length, thick. Tip of spines reddish, in two sizes, large and small, alternating; midrib below unarmed for the lower one third, in three rows; marginal spines at the base 2–2.5 mm long, 11–15 mm apart, almost straight, middle part of leaf 2 mm long spines; 8–9 spines in 10 cm area and crowded towards apex; midrib prickles curved towards apex, at the base 4–6 mm long, 2–7 mm apart, at middle 4–6 mm long 2–3 cm apart and crowded towards apex.

Inflorescence terminal, compound spike; up to 40 cm long, up to 14 cm wide; bracts ~15, yellowish, lowest one 36 cm long, 7 cm wide, lanceolate, margin and midrib of the bracts are prickled near the tip; median bract 17 cm long, 3 cm wide, lanceolate, moderately fragrant; spikes numerous, about 15 in number, and 5–12 cm long, 13–15 mm in diameter, cylindric, dense; staminal fascicles 7–10 mm long, of 6–10 stamens; column 4–6 mm long, slender; at apex branching umbellately; free filament tips 1.5–3 mm long; anthers 2.3–2.7 mm long, oblong, bearing at apex a 0.1–0.2 mm subulate prolongation of the connective.; staminate spike 6–15 × 2 cm, cylindrical; androphore 0.6–0.8 cm long, stamenophore columnar, 0.3–0.4 cm long with 7–9 stamens; stamens in phalanges, 0.3–0.4 cm long, crowded at the apex of stemonophore; filaments 0.1 cm long, arranged in pairs; anther 0.2–0.3 cm long.

Infructescence terminal, solitary, pendent syncarp, peduncle 60–85 cm long, 2–2.5 cm in diameter, 3-sided; syncarp globose, 20–35 cm in diameter, 52–86 phalanges per syncarp, 9.5–13.5 cm long, 4–6.5 cm wide, 3.6–5 cm thick, 5–6 angled, cuneiform, compressed, 1 fruit per infructescence, the apex of drupe concave and its tip 7–18 mm wide, ¼–1/3 free, the surface glaucous smooth, shiny, the apex sharply 5–6 angled; carpels 1–4, commonly 2–3, in 1 (or 2) parallel lines and the stigmas directed to one side, 1.5–3.5 mm long, elliptic to ovate, brown, papillose, separated by a 0.5–2 mm valley, and behind them on the actual apex is a concavity 1–2 mm deep. Endocarp supramedian to the lower 2/5 and 20–35 mm long, bony, dark brown, the apex truncate or with a conic projection, the lateral walls 2–5 mm thick, within smooth, shiny; seeds 10–16 cm long, broad ellipsoid; when single carpellate the endocarp 16–24 mm long, subglobose, and the seed subglobose; apical mesocarp 4–7 cm long; basal mesocarp 3–5 cm long, fibrous and fleshy; seeds 9–10 mm long, 6.5–7 mm in diameter, obconic (Figs. 4.29 and 4.30) (St. John 1975; Rahayu 2011).

Flowering and fruiting: Not recorded.

Leaf micro-morphology: (Fig. 4.31)
In *P. dubius* papillae are absent. Similar observations were reported by Tomlinson (1965) and Kam (1971). According to Tomlinson (1965) and Kam (1971), *P. dubius* showed Class 1 (unspecialized stomata) type of stomata.

Notes:
Pandanus dubius is grown in South Indian gardens for its fragrant flowers and as an ornamental species. It is native to tropical Asia between Java and the Caroline Islands (http://www.rarepalmseeds.com/pix/PanDubshtml). The leaf apical tips of *P. dubius, P. unguifer,* and *P. martinianus* are abruptly narrowed and bearing a subulate tip. *P. dubius* is clearly distinguished from the other Indian *Pandanus* species in having very broad and thick leaves. It is a tree species having large size prop roots. According to Rahayu (2011), *P. dubius* appears close to *P. bidur*. Morphologically, these species are similar to each other in having prop roots long (4 m), blade linear, leaf coriaceous, leaf apex abruptly terminating at once in a point, tertiary cross vein forming oblong meshes. *Pandanus bidur* differs from *P. dubius* by three character states; phalange clavate, stigma lip-like and supramedian endocarp (Rahayu 2011).

Fig. 4.29 *Pandanus dubius*: (**a**) habit; (**b**) details of leaf size

Specimens examined:
India, Karnataka, Udupi, alt. 12 m, 13°19′50″ N, 74°45′15″ E). *Rahul Zanan* 11

***Pandanus leram* Jones *ex* Fontana,** Asiat. Res. 3: 163, 1792; Karthikyan et al., Fl. India, Florae Indicae Enumeration: Monocotyledonae, 177–178, 1989. *P. andamanensium* Kurz, H, vi, 485; Br, 659.

Expanded diagnosis:
Large tree, about 8–16 m in height, erect, numerous irregularly branched, stem 15–20 cm in diameter, smooth; prop roots 2–4 ft long at the base, smooth, 4–6 cm in diameter; leaf long with sheathing at the base, leaf not glaucous above, linear, 250–350 cm long and 7–9 cm wide, margins and mid vein spiny with sharp marginal and curved mid vein spines, three rows, throughout the leaf, black in color, thick,

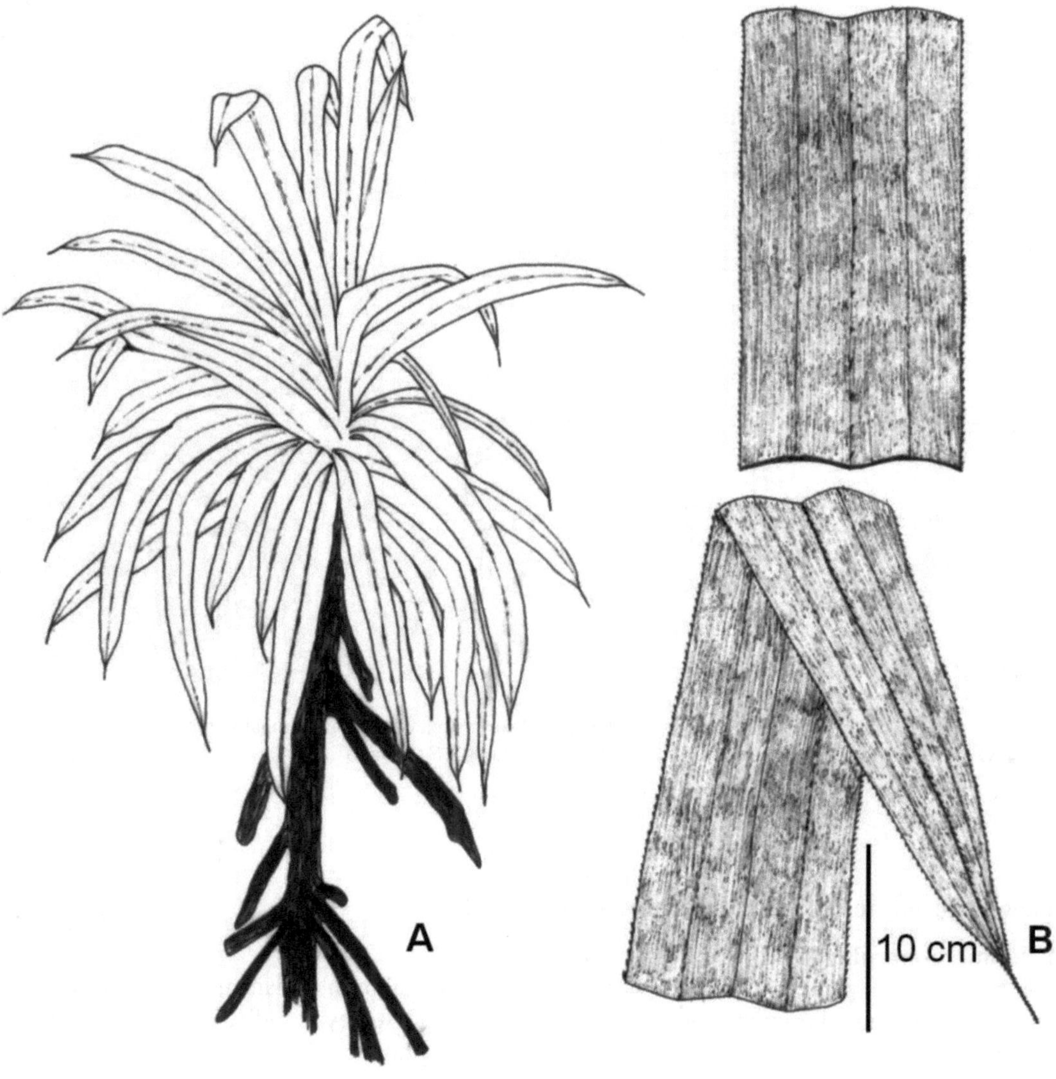

Fig. 4.30 *Pandanus dubius*: (**a**) habit; (**b**) adaxial and abaxial leaf surface

on margin basal spines 4–6 mm long, 20–30 mm distance (4–6 prickles per 10 cm), to the middle part if the leaves midrib spines absent, on middle part marginal spines 2–4 mm long, 0.8–1.7 cm distance (8–9 spines per 10 cm), on midrib spines 2–3 mm long, 2.5–5 cm distance (3–4 spines per 10 cm) and crowded towards apex.

Inflorescence terminal, compound spike; spathe 150–180 cm long with 10–12 yellowish, variable in length, spiney bracts, fragrant; staminate spike cylindrical 8–15 cm long, 3–4 cm wide, androphore 1–1.2 cm long, stemonophore 0.4–0.5 cm long with 12–14 stamens; stamens 0.4–0.5 cm long with short filament; anther 0.3–0.4 cm long.

Infructescence terminal, woody, bracteates, variable length, solitary, ellipsoid or globose, 18–31 cm long, 18–24 cm diameter, drupes connate, subtruncate top, 7–10 cm long, 3–7 cm wide, compressed, hexagonal, low pyramidal, smooth; style very short; stigma 3–14, brownish, distributed over the top of the carpel, usually more than 5 and then placed in 2–3 irregular rows, broad, flat, erecto-patent; endocarp abonic, mesocarp fibrous (Stone 1975; Rahayu 2011) (Figs. 4.32 and 4.33).

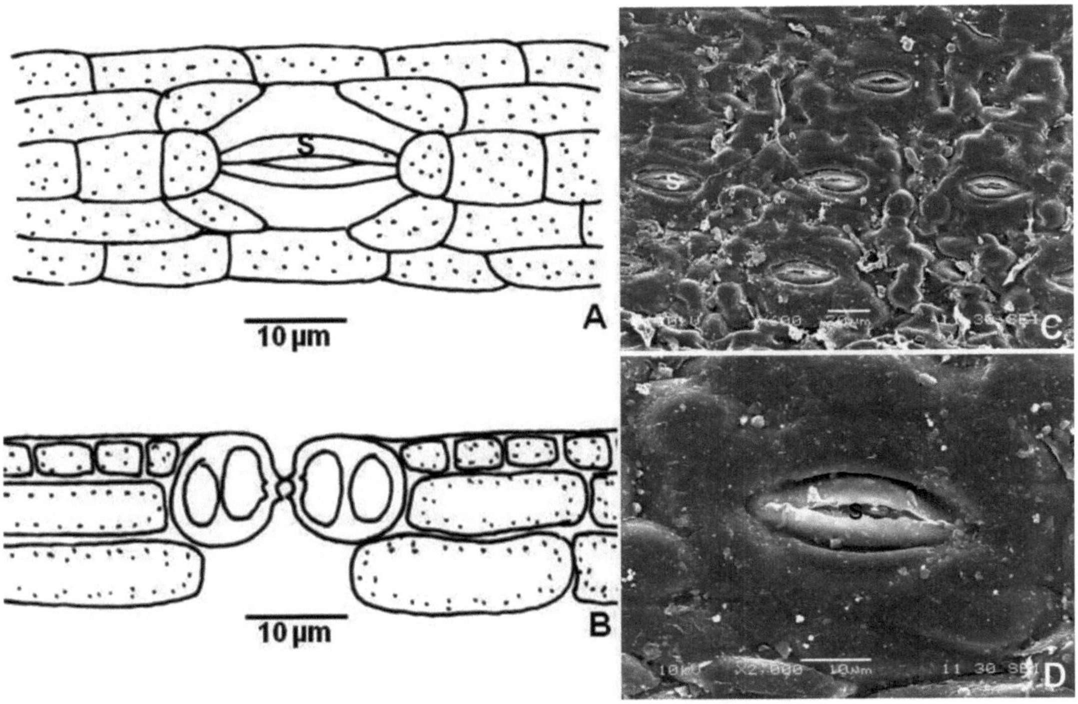

Fig. 4.31 Abaxial leaf epidermal papillae in *P. dubius*: (**a**) surface view; (**b**) longitudinal section of abaxial epidermis (**c**) SEM of papillae (×600); (**d**) close view around stoma (×2,000) (*S* stomata)

Flowering: July to October, **Fruiting:** August to March.

Leaf micro-morphology: (Figs. 4.34 and 4.35)
For the analysis of papillae in *P. leram* we received leaf material from Andaman Island and we personally collected material from the botanical garden of Botanical Survey of India, Yarcud, Salem district, Tamil Nadu. In the former locality, we could not record the presence of papillae on abaxial epidermis, whereas the leaf material from the latter locality showed the presence of papillae. As mentioned elsewhere, in this locality, the species is cultivated in the botanical garden. The micro-morphological details are as follows. The abaxial epidermal papillae are randomly distributed, long elongated, unbranched, dome shaped, and separately borne (15–20 µm long); some papillae are randomly attached to each other, forming a row (3–4 papillae) that is parallel to cell length. The terminal subsidiary cells papillae are elongated, unbranched, tending to overarch guard cells of stomata (18–22 µm long). The lateral subsidiary cells papillae are unbranched, dome shaped with 3–5 small papillae (10–15 µm long). According to Tomlinson (1965) and Kam (1971), Class 4 (papillose neighboring and subsidiary cells) type of stomata.

Notes:
Pandanus leram represents subgenus Rykia, section *Hombronia*. In the Andaman and Nicobar Islands, *P. leram* is evidently present in wild form, with larger phalanges. The cultivated species from Tamil Nadu is only the male specimen and matches with the Andaman Nicobar populations. *P. leram* appears to be in close relation with *P. dubius*. Morphologically, these species are similar to each other in having linear, coriaceous leaves, leaf apex abruptly terminating at once in a point. *P. dubius* differs from *P. leram* by three character states, tertiary cross veins form oblong meshes, phalange cuneiform,

Fig. 4.32 *Pandanus leram*: (**a**) habit; (**b**) staminate inflorescence; (**c**) stamens

and stigma cordate to reniform (Rahayu 2011). In the Andaman and Nicobar Islands, fruits are used as staple food, eaten after boiling and processing; the leaf used for thatching roof, dried fruit with fiber is used for cleaning the hands.

Specimens examined:
INDIA, Tamilnadu, Salem District, Yarcud, alt. 1,400 m, 11°47′06″ N, 78°13′25″ E, *Rahul Zanan* 14 (steminate); same locality, *Rahul Zanan* 26 (steminate), Andaman and Nicobar Island *Rahul Zanan* 59 (pestilate).

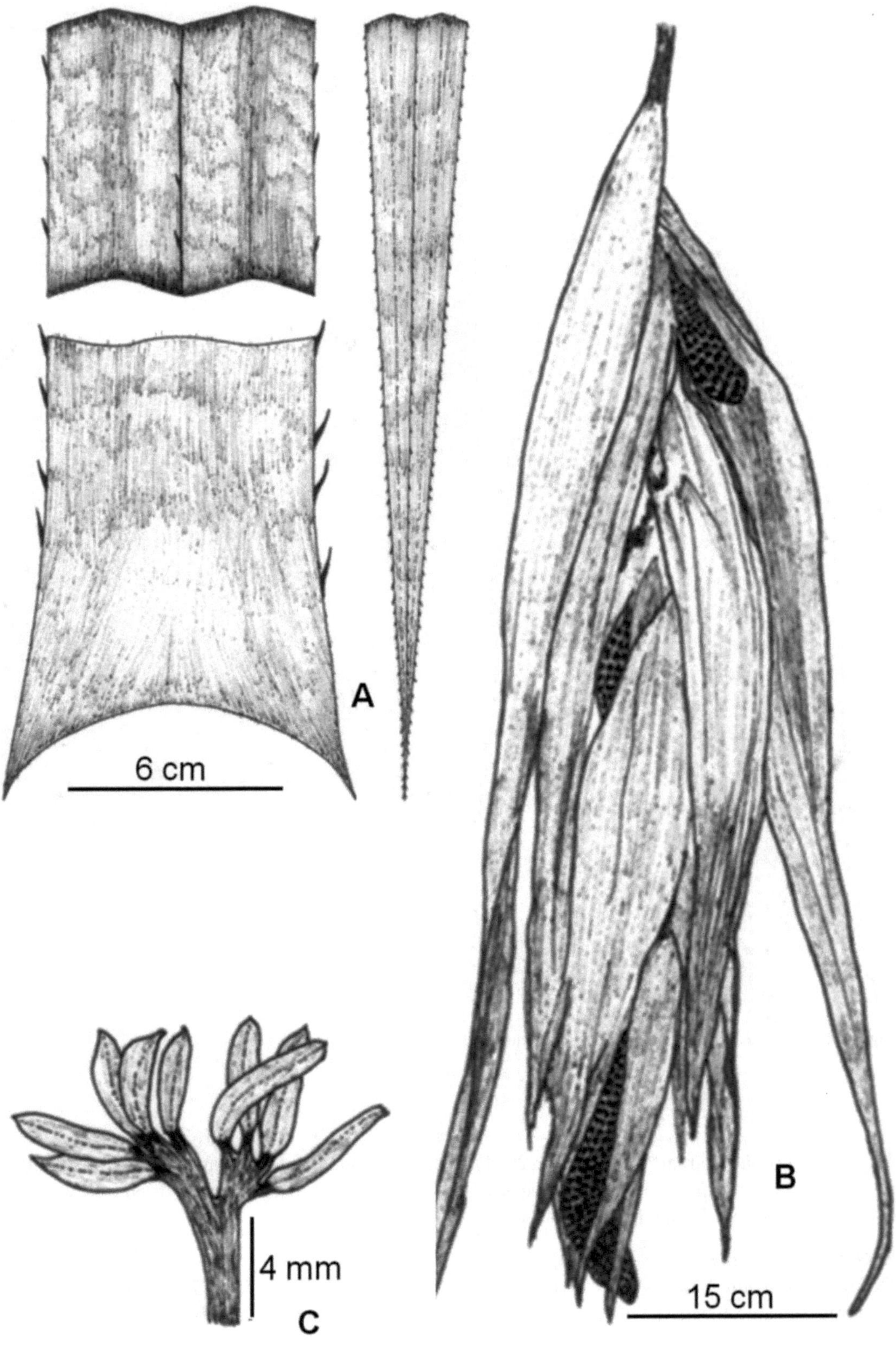

Fig. 4.33 *Pandanus leram*: (**a**) adaxial and abaxial leaf surface; (**b**) staminate inflorescence; (**c**) androphore

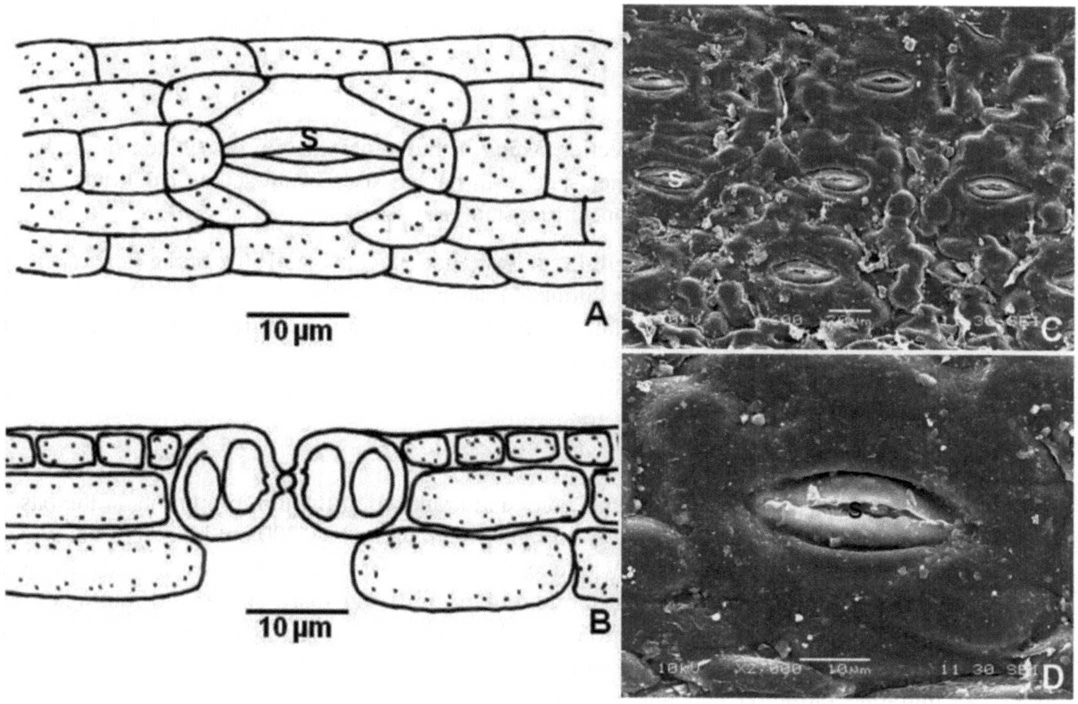

Fig. 4.34 Abaxial leaf epidermal papillae in *P. leram*: (**a**) surface view; (**b**) longitudinal section of abaxial epidermis; (**c**) SEM of epidermis (×600); (**d**) close view around stoma (×2,000) (*S* stomata)

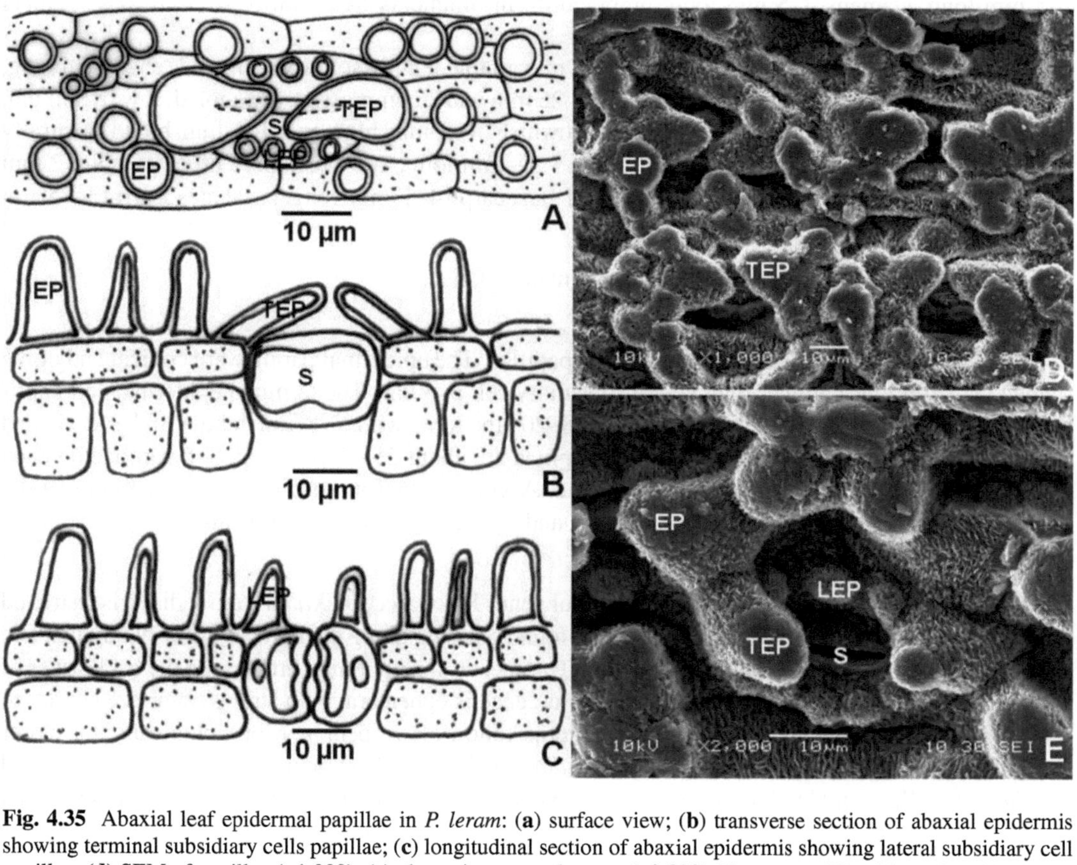

Fig. 4.35 Abaxial leaf epidermal papillae in *P. leram*: (**a**) surface view; (**b**) transverse section of abaxial epidermis showing terminal subsidiary cells papillae; (**c**) longitudinal section of abaxial epidermis showing lateral subsidiary cell papillae; (**d**) SEM of papillae (×1,000); (**e**) close view around stoma (×2,000) (*S* stomata, *EP* epidermal papillae, *TSP* terminal subsidiary cell papillae, *LSP* lateral subsidiary cell papillae)

Section *Kaida* St. John Ms.

The single species *P. kaida* representing section *Kaida* in India; abundantly distributed in the west coast of India. It is distinguished from other species in having style ovate-acute or deltoid; carpel medium, free or connate in phalanges or concentric; stamens umbellate.

Pandanus kaida Kurz, J. Asiat. Soc. Bengal, Pt. 2, Nat. Hist. 38 (2): 148. 1869; Stone in Dassan. & Fosb. Rev. Handb. Fl. Ceyl, 3: 308, 1981; Stone, Malayan Agriculturist 9: 34–44. 1970; Trimen, Handb. Fl. Ceylon 4: 341, 1898; Stone in Matthew, Fl. Tamilnadu Carnatic 3: 1683. 1983; Karthikyan et al., Fl. India, Florae Indicae Enumeration: Monocotyledonae, 177–178, 1989; Bhat in Fl. Udupi, 670–674, 2003. *P. candelabrum* sensu Kurz in J. Bot. 5:127, 1867, non Beauvois. *P. siamensis* Williams, Bull. Herb. Bioss. Ser. II, 4: 220, 1904.

Expanded diagnosis:
A large shrub or small tree, 1–4 m tall; stems erect and branched at the top, prop roots at the base, 2–3 ft long. Leaves linear-ensiform with deep to light green, up to 180–200 cm long, 4–5 cm wide, acute apex, margin and midrib spiny with whole leaf, spines pale white in three rows, midvein with sharp curved prickles; Spines on midrib at the base 4–5 mm long, sharp and 35–40 mm apart, at middle spines shorter, 3–4 mm long, close to each other and crowded towards apex. Similarly, marginal spines at base, sharp, 15–22 mm apart with 3–5 mm long, and at middle position marginal spines shorter, 2–3 mm long, close to each other and crowded towards apex.

Inflorescence terminal, ephemeral in nature with strong fragrance, 10–12 yellowish bracts with variable length, margin and midrib with prickles; linear-lanceolate or lanceolate consisting of dense spikes; spikes 8–12 cm long, cylindrical, stamens 17–23, androphore 9–10 mm long, *stemonophore* 4–5 mm long, stamens 4–5 mm long, umbraculiform, filaments free with 0.5–2 mm long; anthers linear, 3–4 mm long.

Infructescence terminal, woody, bracteates with variable length, solitary, ellipsoid, 25–30 cm long, 16–20 cm diameter, drupes 40–50 mm long, 20–25 mm wide, compressed, hexagonal, low pyramidal, smooth; style very short or absent; stigma, brownish, 1–2 mm high and 1–3 mm broad, centrally grooved, facing each other; endocarp abonic, 15–18 mm long, 1–5 seeds per carpel with 9–12 mm long, 2–3 mm diameter; mesocarp fibrous, apical mesocarp 14–17 mm long, basal mesocarp 13–15 mm long (Figs. 4.36 and 4.37).

Flowering: July to November **Fruiting:** September to January.

Leaf micro-morphology: (Fig. 4.38)
In *P. kaida,* the abaxial epidermis is covered by papillae. The epidermal papillae are conical and dome shaped (7–8 μm long), longer than lateral subsidiary cell papillae. The terminal epidermal cell papillae are long, distinguishable in to bilobed (18–20 μm long with 7–8 μm lobes) and rod shaped (9–11 μm long). The lateral subsidiary cells papillae are about 25–27 μm long, 3–6 papillae on each side, forming a wall surrounding to the stomata. According to Tomlinson (1965) and Kam (1971), *P. kaida* showed Class 4 (papillose neighboring and subsidiary cells) type of stomata.

Notes:
Pandanus kaida is the only species representing subgenus Rykia, section *Kaida* from India. It is restricted to the west coast of the Malabar region of India; found growing along the borders of the paddy fields in Goa, Karnataka, Kerala, and Tamil Nadu states. The species mostly found growing in the back water of sea or lowland areas of costal region. Male inflorescence is ephemeral with strong fragrance similar to *P. odorifer.* Mature fruit dark green in color and turns yellow after ripening. Morphologically it is closer to *P. odorifer* (subgenus Pandanus, section *Pandanus*), Stone (1974) mentioned that Martelli placed *P. kaida* under Section *Hombronia* that was attributed to the Section Pandanus. In 1981, Stone shifted

Fig. 4.36 *Pandanus kaida*: (**a**) habit; (**b**) staminate inflorescence; (**c**) stamens; (**d**) syncarp; (**e**) stigmas in close view

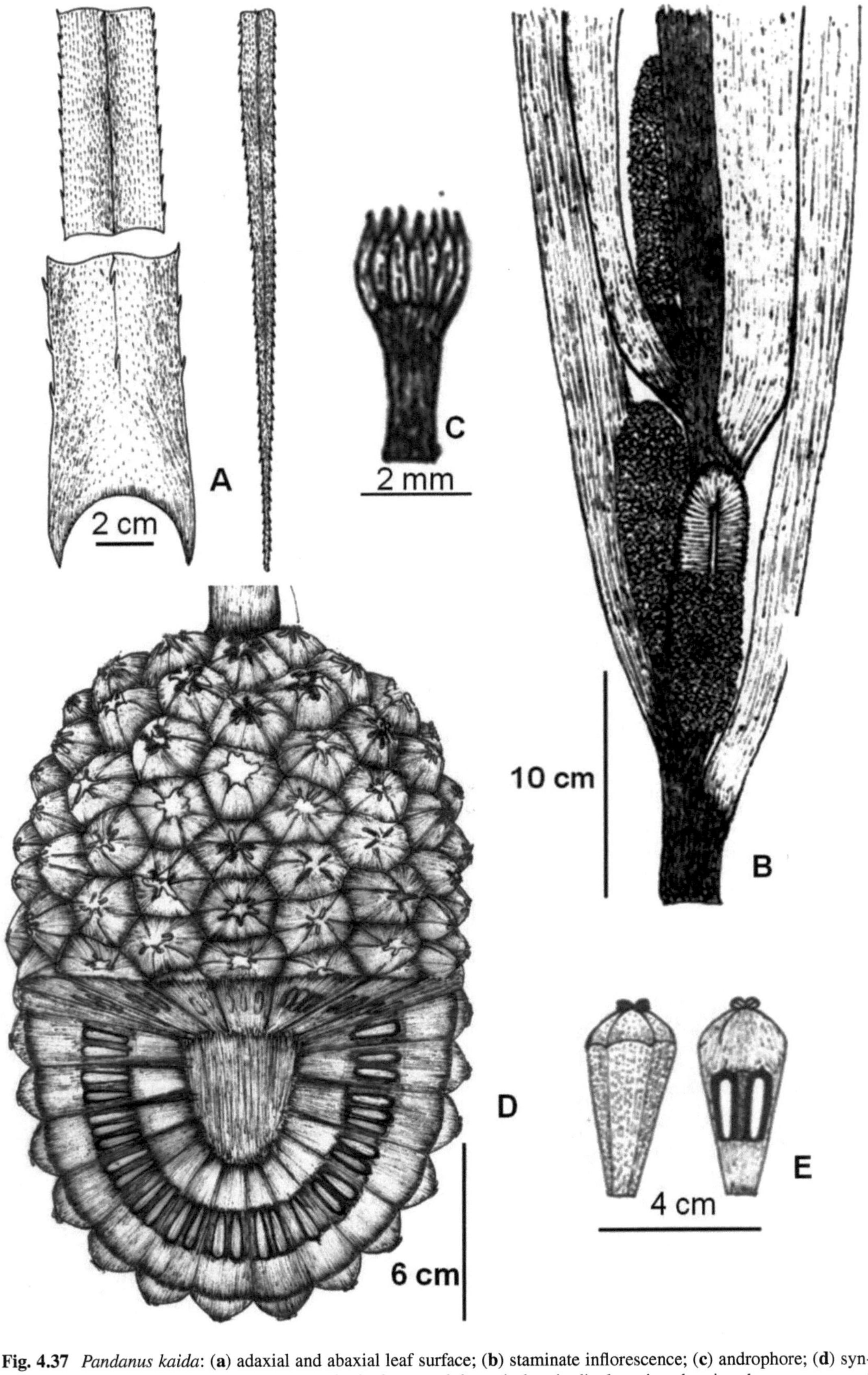

Fig. 4.37 *Pandanus kaida*: (**a**) adaxial and abaxial leaf surface; (**b**) staminate inflorescence; (**c**) androphore; (**d**) syncarp, *lower* half in longitudinal section; (**e**) single drupe and drupe in longitudinal section showing the mesocarp and endocarp

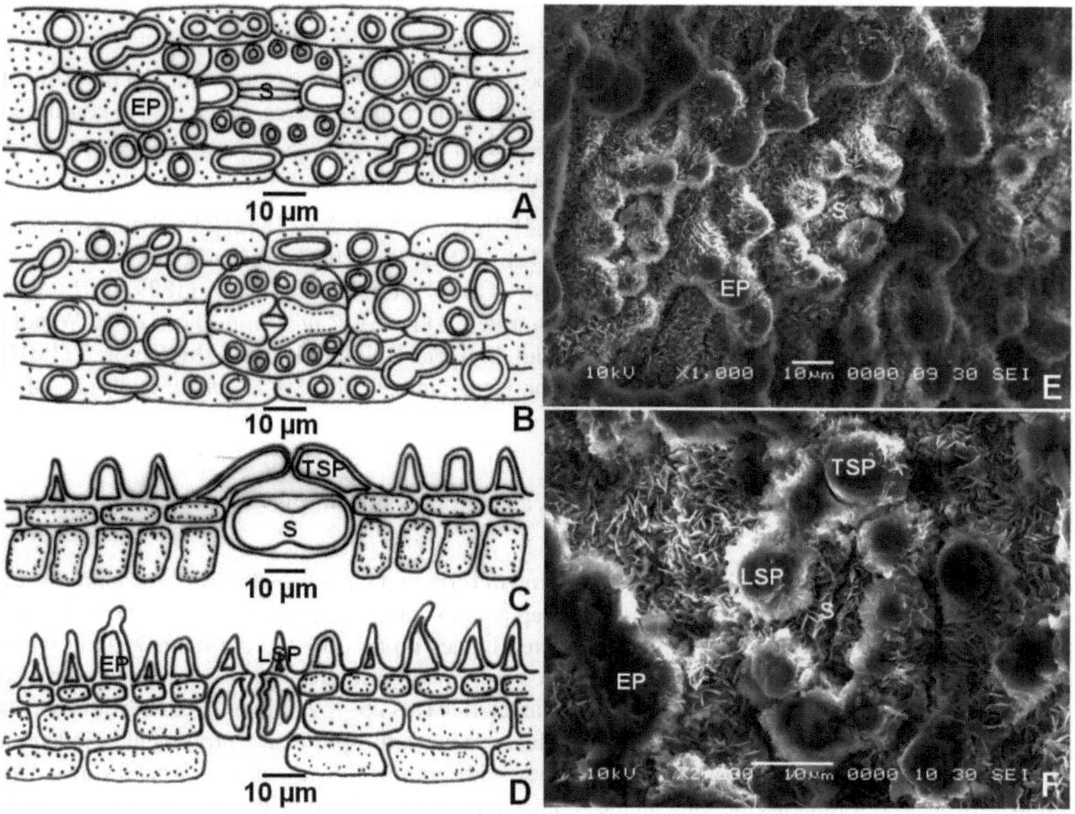

Fig. 4.38 Abaxial leaf epidermal papillae in *P. kaida*: (**a**) surface view showing *rod shaped* terminal subsidiary cells papillae; (**b**) surface view showing branched terminal subsidiary cells papillae; (**c**) transverse section of abaxial epidermis showing terminal subsidiary cells papillae; (**d**) longitudinal section of abaxial epidermis showing lateral subsidiary cell papillae; (**e**) SEM of papillae (×1,000); (**f**) close view around stoma (×2,300) (*S* stomata, *EP* epidermal papillae, *TSP* terminal subsidiary cell papillae, *LSP* lateral subsidiary cell papillae)

P. kaida from Section Pandanus to Section Kaida. Thus, its position before 1981 was in Section Pandanus and our analysis supports this placement

Specimens examined:
INDIA, Karnataka, Bhatkal District, Bhatkal, alt. 10 m, 14°04′05″ N, 74°30′16″ E, Rahul Zanan 3 (pistilate); same locality, Rahul Zanan 8 (staminate); Kerala, Alappuzha District, Alappuzha, alt. 7 m, 09°30′16″ N, 76°19′01″ E, Rahul Zanan 63 (pistilate); same locality, Rahul Zanan 64 (staminate); Goa, North Goa district, 3–4 km from Mudgaon towards Cancona, alt. 15°11′23″ N, 73°59′50″ E, Rahul Zanan 25 (pistilate); same locality, Rahul Zanan 26 (staminate).

Subgenus: Kurzia B.C. Stone

The subgenus Kurzia is recognized by the combination of following characters: leaf apex with armed ventral plate, stamens attached in phalanges.

Section *Jenneretia* (Gaudich.) B.C. Stone

Section *Jenneretia* represents a single cultivated species, *P. amaryllifolius,* from India. On the basis of the staminate inflorescence characters and macroscopic features of the leaves, Stone (1978) confirmed the placement of *P. amaryllifolius* under subgenus Kurzia. In this species, because infructescence is not recorded, Stone (1978) tentatively placed it under section *Jeanneretia.*

Pandanus amaryllifolius Roxb. Hort. Bengal (1814) 71; F1. Ind. ed. 2, 3 (1832) 743. Syn.: *Pandanus odoratissimus* Blume, Catalogus, (1823) 111, non L.f. (1782); *P. amaryllidifolius* Voigt , Syll. Ratisb. vol. 2, (1828) 52; *P. amarylloides* Parment. ex Desf. , Cat. pl. hort. Par. ed. vol. 3, (1829) 9; *P. latifolius* Hassk. in Flora vol. 25, (1842) 2. Beibl., 13, non Perrot (1825); *P. moschatus* hort. ex H. Wendl., Ind. Palm., (1854) 45, non Miq. (1855); *P. laevis* Ridl. in Agric. Bull. Straits and F. M. S. vol. 1, (1902) 336, non Kunth (1841); *P. hasskarlii* Merr. , Interpr. Herb. amb., (1917) 80; *P. odorus* Ridl. , Fl. Mal. Penins. vol. 5, (1925) 81.

Original diagnosis:
Diffuse. Leaves linear, tending to be three-nerved, apices rather broad, somewhat spinous-serrulate.

Extended diagnosis:
Small herb grows in two forms recorded: small growth form and large growth form. Aerial roots are distinct in both growth forms, with distinct root cap at apex; large or tall growth form showed erect, branched stem, about 2–4.5 m tall, 15 cm diameter, blades up to 150–220 cm long, 7–9 cm wide, rarely with 1–3 small stout spines on midrib near the base; small growth form showed erect, unbranched slender stem, height of about 1.6 m tall, 2–5 cm diameter, leaves 25–75 cm long, 2–5 cm wide; leaves broad linear, middle or rather dark green above, glaucous beneath, apex acute, prominent twin lateral pleats above, the margins entire except at leaf apex with presence of few minute prickles less than 1 mm long. The odor remains the same in the small and large growth forms leaves. Prop roots distinct, numerous, long at the base to middle part of the stem with distinct root cap at apex.

Infructescence unknown (never produced (Stone 1978)); inflorescence very rare, once collected and described from a collection from Laguna, Ternate, in the Moluccas, Beguin 1690 (Stone 1974); probably pendent about 60 cm long, white, spathes 90 cm long with several oblong spikes about 35 cm long or more; many crowded staminal phalanges with 4–9 mm long or longer, flat, 1.5–2.5 mm wide, filaments very short, 0.5–1.5 mm long, 0.4–0.6 mm wide; anther oblong, 2.5 mm long, 0.5 mm wide, apex bluntly convex, 3–6 stamens per phalange. *P. amaryllifolius* follows only vegetative mode of reproduction through lateral buds (Wakte 2010) (Figs. 4.39 and 4.40).

Flowering and fruiting: unknown (never produced).

Leaf micro-morphology: (Fig. 4.41)
In *P. amaryllifolius,* the abaxial epidermis is covered by papillae. The epidermal papillae are unbranched, unambiguously distributed, systematically arranged, simple papillae. Each epidermal cell papillae are uniformly distributed in rows of 1–7 papillae (7–8 μm long). The terminal subsidiary cells papillae are elongated, partially covered guard cells (20–21 μm long) of stomata with an average ranging from 12 to 13 μm long. The lateral subsidiary cells are unbranched, 3–7 papillae, 7–9 μm long, producing wall around the guard cells, forming a beautiful, necklace-like structure around the stomata. According to Tomlinson (1965) and Kam (1971), abaxial epidermal papillae of *P. amaryllifolius* showed Class 4 (papillose neighboring and subsidiary cells) type of stomata. Kam (1971) and Wakte et al. (2007) observed similar epidermis in *P. odorus* (synonym of *P. amaryllifolius*).

Notes:
In the infrageneric classification system of Stone (1974), placement of *P. amaryllifolius* is under the subgenus Kurzia Stone and most likely in Section Jeanneretia (Gaudich.) Stone on the basis of the characters of the staminate inflorescence, the structure of the staminal phalanges, anthers and leaves.

Fig. 4.39 *Pandanus amaryllifolius*

Section Kurzia Stone and Microstigma Kurz. are two other possible sections where this species may find its place (Stone 1978). *P. amaryllifolius* is usually propagated through rooted suckers or stem cuttings. *P. amaryllifolius* is the only species of family Pandanaceae with fragrant leaves (Wakte et al. 2009) and it is cultivated in tropical peninsular countries for flavoring various foods such as bakery products, sweets, and even home cooking (Jiang 1999; Laohakunjit and Noomhorm 2004; Bhattacharjee et al. 2005; Wakte et al. 2007, 2009). Traditionally, in India, leaves have been being used in cooking rice to impart an aroma resembling costly scented rice such as Basmati. In Karnataka and Tamil Nadu states, *P. amaryllifolius* is commercially cultivated and row leaves are sold in vegetable markets (Wakte et al. 2009).

Specimens examined:
INDIA. Karnataka, North Canada district, Avarsa, alt. 12 m, 14°43′16″ N, 74°16′47″ E, *Rahul Zanan* 12; India, Karnataka, Udupi district, Udupi, alt. 12 m, 13°19′50″ N, 74°45′15″ E, *Rahul Zanan* 39.

Subgenus Pandanus

The subgenus Pandanus is recognized by the following combination of characters: carpels connate into phalanges, irregular shape, individual carpel apices convex, separated by deep grooves; stigma V-shaped; stamens connate into phalanges.

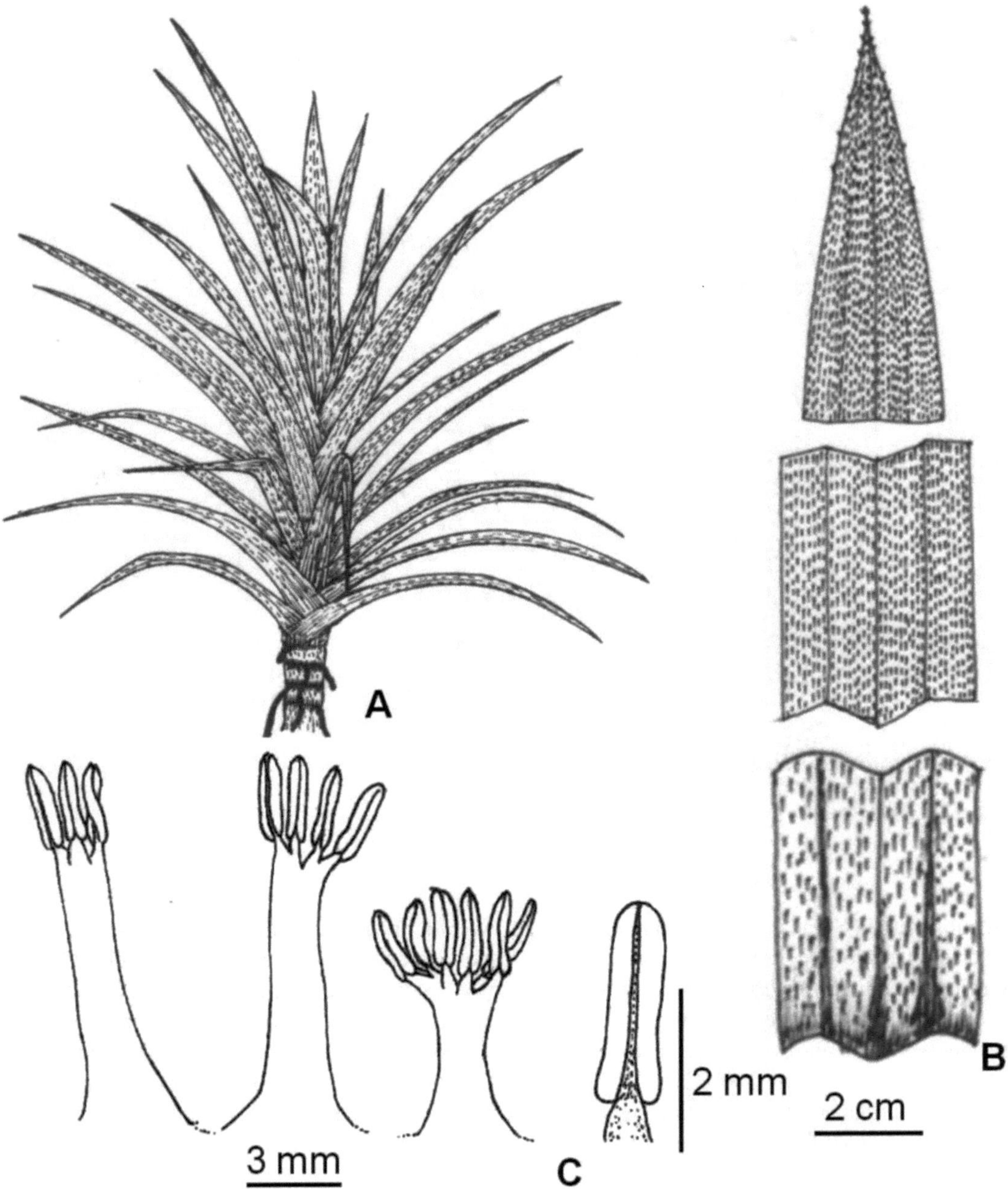

Fig. 4.40 *Pandanus amaryllifolius*: (**a**) habit; (**b**) adaxial and abaxial leaf surface; (**c**) Stamens (Redrawn from Stone (1978)

Section *Pandanus*, Subsection Pandanus

This section is polytypic. *Pandanus odorifer* belongs to subsection Pandanus; a well-known and widely distributed species mostly in costal regions and rarely in landward regions. It is recognized by the combination of following characters: carpels of each phalange arranged concentrically, phalanges more than 20, phalanges apex convex to subtruncate; stigma predominantly horizontal; staminal phalanges racemose.

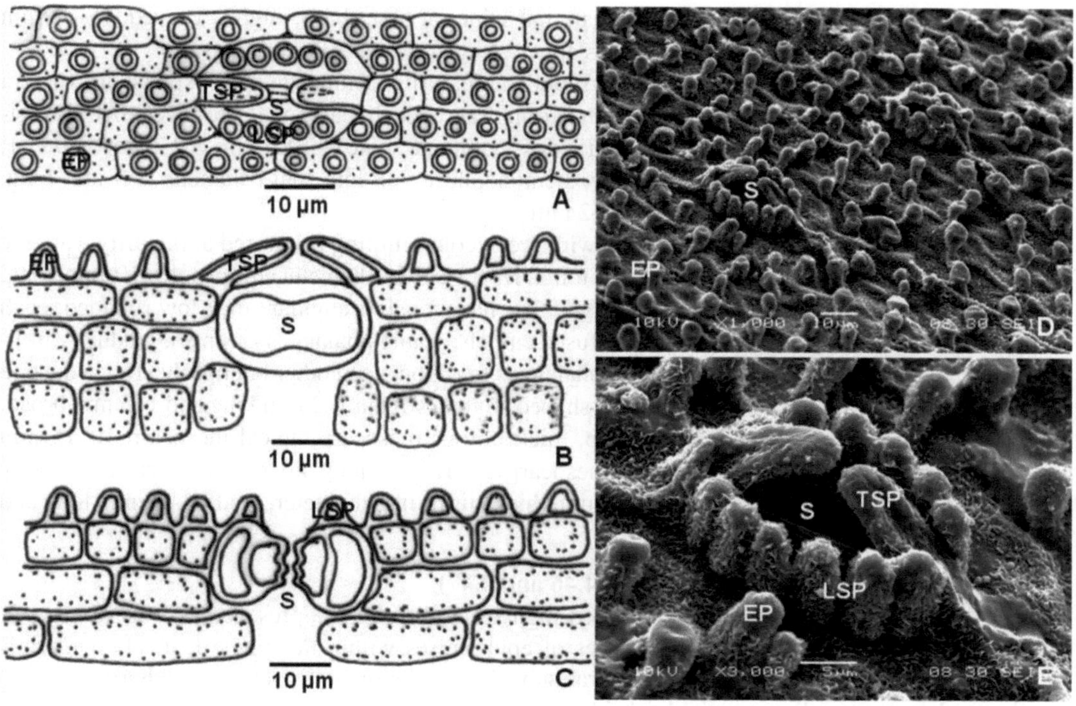

Fig. 4.41 Abaxial leaf epidermal papillae in *P. amaryllifolius*: (**a**) surface view; (**b**) transverse section of abaxial epidermis showing terminal subsidiary cells papillae; (**c**) longitudinal section of abaxial epidermis showing lateral subsidiary cell papillae; (**d**) SEM of papillae (×1,000); (**e**) close view around stoma (×3,000) (*S* stomata, *EP* epidermal papillae, *TSP* terminal subsidiary cell papillae, *LSP* lateral subsidiary cell papillae)

Pandanus odorifer (Forssk.) Kuntze, Revis. Gen. Pl. 2: 737, 1891. *Keura odorifera* Forssk., Fl. Ageypt.- Arab. 172, 1775. *Pandanus odoratissimus* L. *f.*, Suppl. Pl. 424, 1782; St. John in Taxon 12: 201, 1963; Karthik et al., Fl. Ind. Enum. Monocot. 178, 1989. *P. fascicularis* Lam., Encycl. 1: 372, t. 1, 1785; Hook. *f.*, Fl. Brit. India 6: 485, 1893; C. D. K. Cook, Aquat. Wetl. Pl. India 280, fig. 289 e-g, 1996. *P. tectorius auct. non* Parkinson, 1774; T. Cooke, Fl. Bombay 3: 324, 1967 (Repr.); R. S. Rao, Fl. Goa 2: 451, 1986; H. B. Naithani et al., For. F. Goa 610, 1997.

Original diagnosis:

Pandanus Rumpf (1744) IV 139–154 t. 74–81, excl. 153 t. 82, = *Keura* Forsk. 1775 = *Arthrodactylis* Forst. 1776. Es wird Pandanus in der Regel L.f. (1781) zugeschrieben; wollte man das aber gelten lassen, also Rumpf's von Linnaeus pater nur in der Dissertatio herb. amb. als genus obsoletum (!) 1753 erwähnten und sonst todtgeschwiegenen Pandanus bei Seite schieben, so müsste der Name Pandanus überhaupt zu den Synonymen gestellt werden, weil 2 andere Namen, also in erster Reihe *Keura* dann die Priorität haben würden. Es ist aber nicht gerechtfertigt, die Runipf-Burniann'schen Namen au verwerfen, t. 82, die L.f. schon aus dem Rumpfsehen Ci tat Hess, gehört zur nächstverwandten Gattung Freycinetia. Rumpf beschreibt übrigens diese kletternde Gruppe für sich und hatte zwischen derselben und seiner 1. Pandanus-Gattung mit 8 Arten eine andere Gattung eingeschoben. P. odorifer OK. (*Keura odoriferus* Forsk. 1775 = *P. odoratissimus* L.f. 1781).

Expended diagnosis:

Large shrubs or small tree, about 9 m tall; stem erect or more or less decumbent, much branched, rough, branching dichotomous or irregular; prop roots numerous, very long at the base to middle; leaves linear-ensiform with deep green, long, up to 2 m × 4–7 cm, acute apex, margin and midvein spiny with whole leaf, spines yellow in three rows, midvein with sharp and curved spines at the base; spines on midrib at the base 5–7 mm long, sharp and 30–40 mm apart, at middle spines close to each other and crowded towards apex; similarly marginal spines at base sharp, slightly curved, 15–25 mm apart with 5–7 mm long and crowded towards apex.

Inflorescence terminal, ephemeral, decays rapidly at maturity, composite spike; spathe 80–100 cm long, 11–12 yellow bracts with variable length, strong fragrance, linear lanceolate-lanceolate consisting of dense spikes, margin and midrib with prickles; staminate spike 6–10 cm long, 2–3 cm diameter; stamens 19–23, androphore 11–12 mm long, stemonophore 7–8 mm long, stamens 3–5 mm long, anther 3–4 mm long, filaments free with 0.5–2 mm long.

Pistilate inflorescence terminal, bracteate with green color, turning bright red at maturity, variable in length, woody, oblong round, 25–30 cm long, 18–25 cm in diameter, drupes 40–50 mm long, 25–35 mm wide, compressed, smooth, pentagonal or hexagonal or angular, apex low rounded; each drupe contains 5–15 concentrically arranged fused carpels, young phalanges greenish, gradually turning yellow and orange-red color at maturity, phalanges free from each other but tightly crowded; style absent; each carpel contains single 'U' or 'V' shaped stigma, stigma 1–2 mm long and 1–3 mm broad, brown; each phalanges showed thin pericarp, thick fibrous mesocarp, apical mesocarp 10–15 mm high with 4–5 mm high pileus and basal mesocarp 13–15 mm long fibers; endocarp thin woody 16–20 mm long, radish brown; seed coat 4–5 mm thick with 1 mm diameter, seed 10–12 mm long and 2–3 mm diameter (Figs. 4.42 and 4.43).

Leaf micro-morphology: (Figs. 4.44, 4.45, 4.46 and 4.47)
Pandanus odorifer abaxial epidermis recorded high variability for the presence and distribution of papillae. We came across four different types of epidermises within the single leaf: (1) epidermis without papillae; (2) papillae on single lateral subsidiary cells or both lateral subsidiary cells; (3) papillae on lateral as well as terminal subsidiary cells; and (4) papillae found on stomatal cells and adjacent epidermal cells. The terminal subsidiary cells papillae are prominent, fully developed, and partially overarch the stomatal pore (12–15 μm long). The lateral subsidiary cells papillae are small, simple, and globular (6.5–10 μm long), 4–6 papillae are on a single side. Solla (1884), Tomlinson (1965), and Kam (1971) observed similar type of variation in the epidermis of *P. odoratissimus*. According to Tomlinson (1965) and Kam (1971), *P. odorifer* shows Class 1 (unspecialized stomata), Class 2 (papillose lateral subsidiary cells), Class 3 (papillose terminal and lateral subsidiary cells), and Class 4 (papillose neighboring and subsidiary cells) type of stomata.

Flowering: July to October **Fruiting:** Nearly throughout the year.

Notes:
Pandanus odorifer is the only native species representing subgenus Pandanus in India. It is highly polymorphous and wildly distributed along the coast. It also wildly cultivated in the Gunjam District of India as a perfume plant for its fragrant flowers (Panda et al. 2007, 2009). In India, several names have been used including *P. odorifer* (Forssk.) Kuntze, *P. fascicularis* Lam., *P. odoratissimus* L.f., and *P. tectorius* Parkinson. All these names are considered to be synonyms except the latter and the name used should be *P. odorifer* (Callmander, personal communication) for the time being. *Pandanus tectorius* has been often misapplied for Indian native species, e.g. Panda et al. (2009). This species is also sporadically cultivated in India in gardens and is not treated in this book (see Stone 1976, 1983). *Pandanus tectorius* is native to the Pacific Islands. Cultivated varieties of *P. tectorius* often have unarmed leaves such as *P. tectorius* var. *laevis* Warb. *Pandanus odorifer*, on the other hand, can be easily recognized by its white and aggressive prickles on the margins and midrib of leaves blades and the "shoulder" on the apex of this mature fruits (Fig. 4.42e).

Specimens examined:
INDIA, Kerala, Trivandrum, alt. 0 m, 08°30′22″ N, 76°53′23″ E, *Rahul Zanan* 9 (pistillate); same locality, *Rahul Zanan* 50 (staminate); same locality, *Rahul Zanan* 51 (pistillate); Maharashtra, Sawantwadi, alt. 61 m, 15°55′47″ N, 73°46′35″ E, *Rahul Zanan* 10 (staminate); Orisa, Ganjam, alt. 16 m, 19°22′17″ N, 85°01′28″ E, *Rahul Zanan* 52 (staminate); Orisa, Ganjam, alt. 8 m, 19°38′51″ N, 85°09′01″ E, *Rahul Zanan* 53 (pistillate).

Fig. 4.42 *Pandanus odorifer*: (**a**) habit; (**b**) staminate inflorescence; (**c**) stamens; (**d**) syncarp; (**e**) stigmas in close view

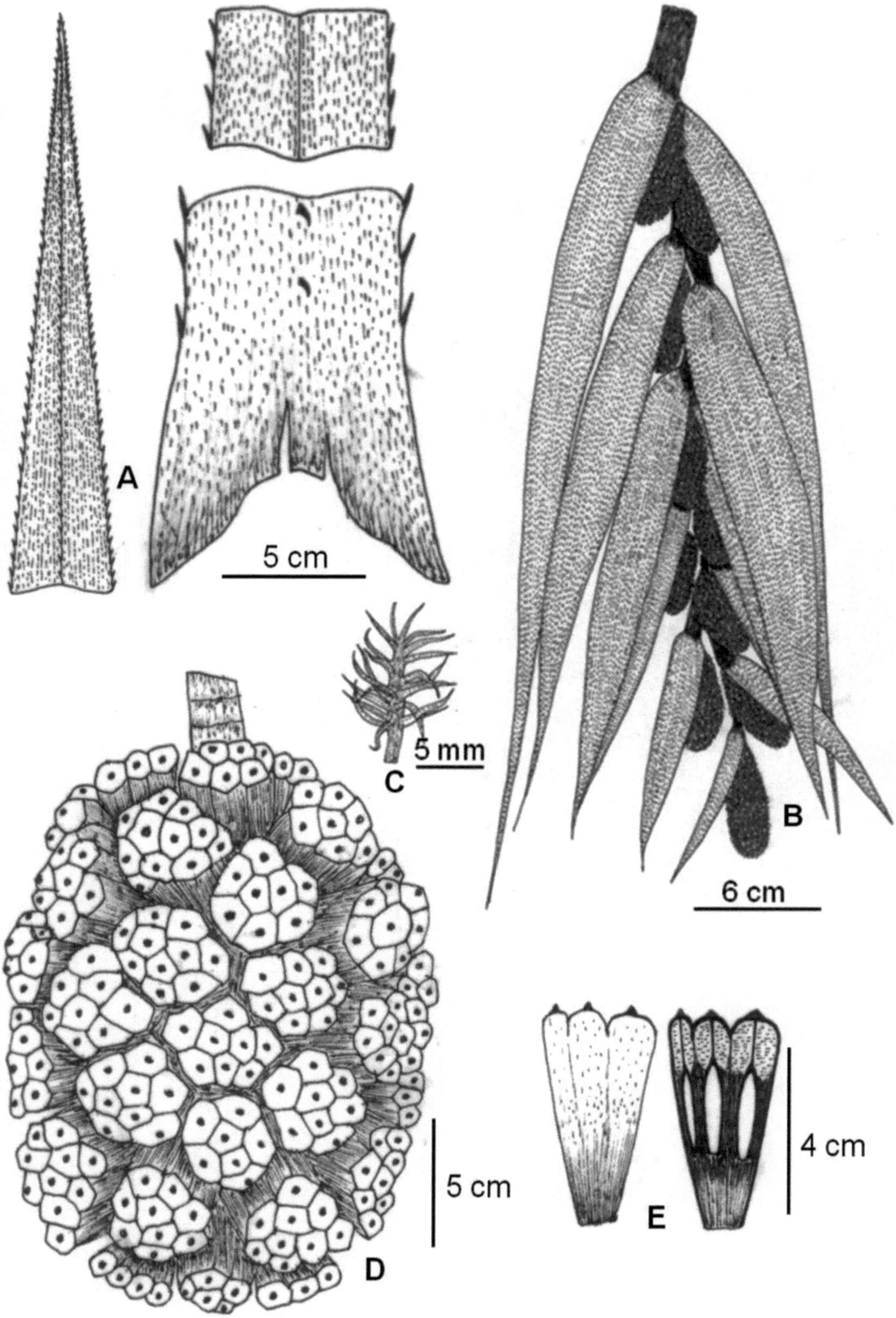

Fig. 4.43 *Pandanus odorifer*: (**a**) adaxial and abaxial leaf surface; (**b**) staminate inflorescence; (**c**) androphore; (**d**) syncarp; (**e**) single drupe and drupe in longitudinal section showing the mesocarp and endocarp

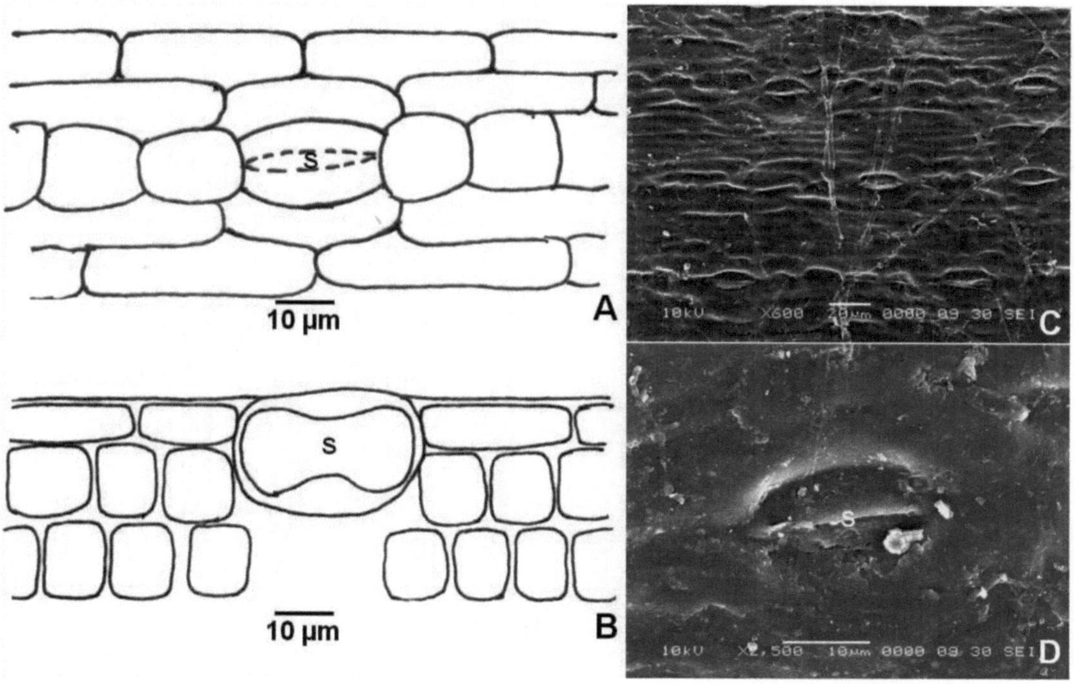

Fig. 4.44 Abaxial leaf epidermal papillae in *P. odorifer*: (**a**) surface view; (**b**) transverse section of abaxial epidermis; (**c**) SEM of abaxial epidermis (×600); (**d**) close view around stoma (×2,500) (*S* stomata)

Benstonea Callm. & Buerki

Benstonea Callm. & Buerki Candollea 67: 328. 2012. *Pandanus* sect. *Acrostigma* Kurz in J. Bot. 5: 100. 1867. *Pandanus* subg. *Acrostigma* (Kurz) B. C. Stone in Bot. Jahrb. Syst. 94: 521. 1974. Type species: *Benstonea affinis* (Kurz) Callm. & Buerki.

Acaulescent or short-stemmed shrubs, often epiphytic, rarely tall trees. Leaves ligulate to linear-attenuate; leaf apex adaxially spinulose along the two main pleats (spines rarely absent). Inflorescences terminal or lateral on short side-branches, pistillate cephalia solitary, sometimes spicately disposed. Cephalia always of simple drupes, drupes never connate into phalanges; pileus usually distinct and calyptrate, grading upward into a hard, spiniform style; stigma linear, always positioned on abaxial side of the style. Endocarp usually with seed-chamber roofed by a thin cartilaginous partition above which is located a small more or less distinct supra-seminal chamber distinct from the rest of the apical mesocarp. Staminate inflorescence normally spicate. Staminate flowers sessile, composed of free stamens with anthers much longer than the short or nearly obsolete filaments, apiculate, sometimes stamens arranged in pauci-staminate dyads, phalanges or triads (Callmander et al. 2012).

This distinct genus is recognized by the combination of following characters: leaf apex with prickly twin plates on the ventral surface, stamens singly attached to spike axis, many crowded, anthers much longer than filaments, cephalia always formed of simple free drupes; style spine-like, slender or erect; stigma linear, on the dorsal side of the style; single-seeded drupes. In India this genus is represented by two species viz., *B. thwaitesii* and *B. foetida*.

Benstonea thwaitesii (Martelli) Callm. & Buerki Candollea 67: 340. 2012. Basionym: *Pandanus thwaitesii* Martelli Webbia 1: 369 (1905); Karthikyan et al., Fl. India, Florae Indicae Enumeration: Monocotyledonae, 177–178, 1989; Bhat in Fl. Udupi, 670–674, 2003; Sharma et al., Fl India, Fl. Karnataka analysis, 295–296, 1984; Gamble, Fl. Pre. Mad. 1094–1095, 1957; Stone, Fl. Cylon Vol III, 293–320, 1981; Henry, et al., Fl. Tamilnadu, Ind., 54, 1989.

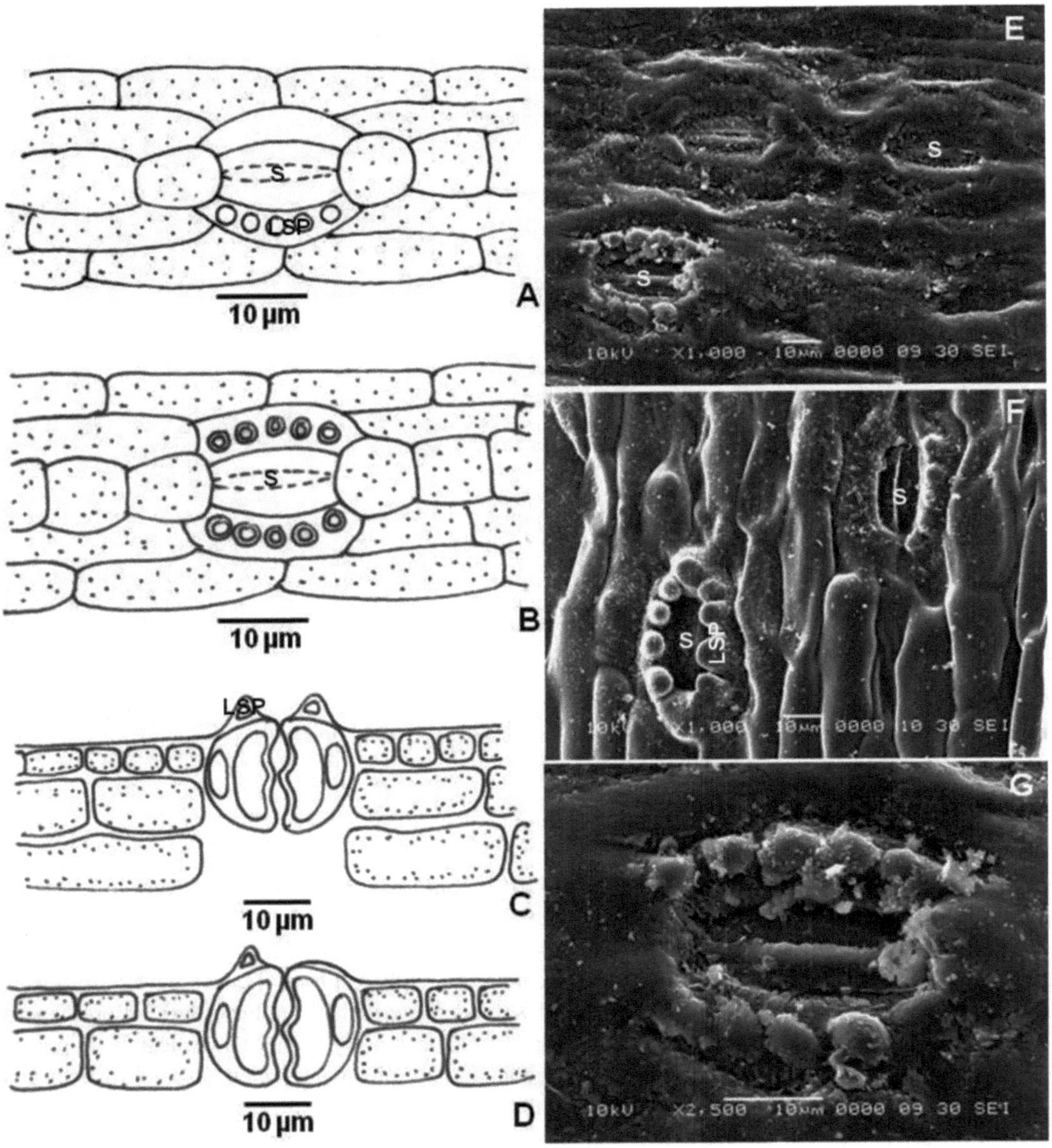

Fig. 4.45 Abaxial leaf epidermal papillae in *P. odorifer*: (**a**) surface view showing papillae on single lateral subsidiary cell; (**b**) surface view showing papillae on both lateral subsidiary cell; (**c**) longitudinal section of abaxial epidermis showing lateral subsidiary cell papillae; (**d**) longitudinal section of abaxial epidermis showing single lateral subsidiary cell papilla; (**e**) and (**f**) SEM of papillae (×1,000); (**g**) close view around stoma (×2,000) (*S* stomata, *LSP* lateral subsidiary cell papillae)

Original diagnosis:

Folina superior rigidia chartacea 3 ½ cent. lata, ultra bimetralia, supra usque ad medium plicato-canaliculata, dein subplana, apicem versus longe attenauato-acuminata, plicis lateralibus, in parte apicali, acutis et tractu circiter 30 cent. longocrebre spinulosis, supra viridian nitentia, venulis transversis numerosissimis brcvissimis, distinctissimis, prominulis notate, subtus glaucescentia minus distinct transverse venulosa et minutissime longitudinaliter striato-nervosa; marginibus spinis pallidis tenuibus armatis; costa media subtus acuta, usque ultra medium spinis reduncis remotiusculis armata. Inflorescentia ♂ composite, elongate; spathis chartaceis lanceolato caeterum spinis numerosis ac minutis sursum versis, armata.

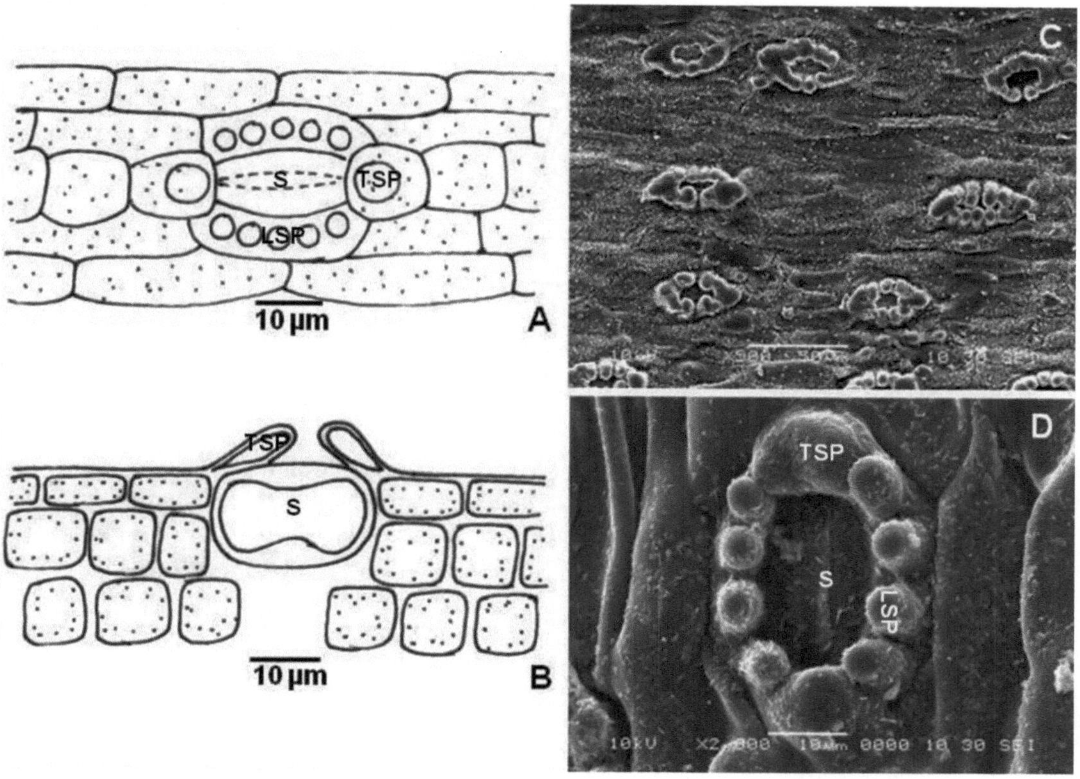

Fig. 4.46 Abaxial leaf epidermal papillae in *P. odorifer*: (**a**) surface view; (**b**) transverse section of abaxial epidermis showing terminal subsidiary cells papillae; (**c**) SEM of papillae (×500); (**d**) close view around stoma (×2,800) (*S* stomata, *TSP* terminal subsidiary cell papillae, *LSP* lateral subsidiary cell papillae)

Pedunculus elongates, 40 cent. longus, digiti minoris crassitiae, trigonus, apice sensim incrassatus, spathis deciduis. Syncarpum solitarium, terminale, magnum, globosum, 22–25 cent. longum, 15 cent. diam. Drupae numerosisimae, elongatolineares, 4 cent. longae, 5 mill. latae (non rite mature) in sicco acute penta-exagonae; pileo indistinct, in parte libera circiter 1 cent. longo, anguste penta-exagono, pyramidato, minute et obscure tuberculato, in stylum longum pyramidatum, subulatum, spiniforme, aduncum vel curvato-adscendentem attenuato; superficie stigmatica deorsum vergenti, lineari in totastyli longitudine nidulanti.

Expanded diagnosis:

Large branched shrub about 3–4 m tall; stem decumbent-ascending or prostate, smooth with a few prop roots on the underground stem; leaves linear up to 360–530 cm long, 4–6 cm wide, margin and midrib spiny with three rows of whole leaf with upper twin plates prickly along distal one-fifth end of the blade; margin with sharp spines and mid veins with curved spines at the base, yellow; at the base 3–4 spines on midrib (per 10 cm) with 2–3 mm long; 15–16 spines on margin (per 10 cm), 2–3 mm long with 7–10 mm distance; at the middle part of leaves 2–3 mm long with 20–30 mm distance on midrib, on margin 7–8 (per 10 cm), 2–3 mm long and crowded towards apex.

Inflorescence terminal, with yellow, very fragrant bracts, variable in length, spinulose-denticulate on the apical margins and killed margins and midrib; distal bracts acute; spike sessile and 8–10 cm long; stamens directly attached to the spike axis not on stemonophore, numerous, 1.7–2.3 cm long with 0.2–0.4 cm long filaments, anthers elongated 2–2.2 cm long.

Carpel simple, ripening as one-seeded drupes (rarely 2–3 seeded), stigma linear on dorsal side of spiniform style; heads racemose-spicate, dorsal placentation with free drupes (Stone 1981). Syncarps up to 6 m long, oblong round, drups 3–3.2×0.5–0.8 cm, connate, slightly narrowed towards base, free apex pentagonal-pyramidal, terminated by the entire subspinescent subulate style (Stone 1981) (Figs. 4.48 and 4.49).

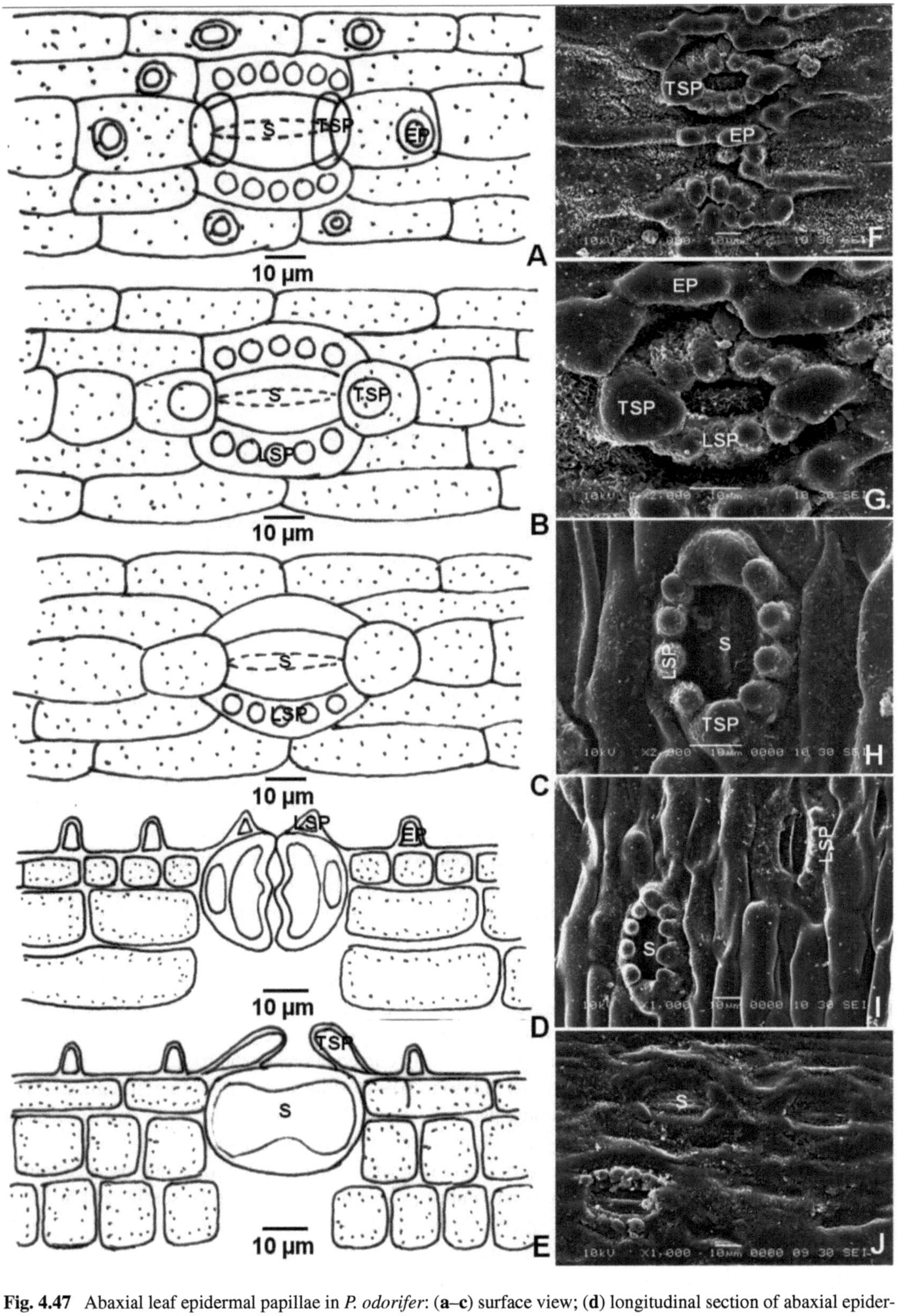

Fig. 4.47 Abaxial leaf epidermal papillae in *P. odorifer*: (**a–c**) surface view; (**d**) longitudinal section of abaxial epidermis showing lateral subsidiary cell papillae; (**e**) transverse section of abaxial epidermis showing terminal subsidiary cells papillae; (**f**) SEM of papillae (×1,000); (**g** and **h**) close view around stoma (×2,000); (**i** and **j**) SEM of papillae (×1,000) (*S* stomata, *EP* epidermal papillae, *TSP* terminal subsidiary cell papillae, *LSP* lateral subsidiary cell papillae)

Fig. 4.48 *Benstonea thwaitesii*: (**a**) habit; (**b**) leaf measuring ~6 m in length; (**c**) staminate inflorescence; (**d**) stamens

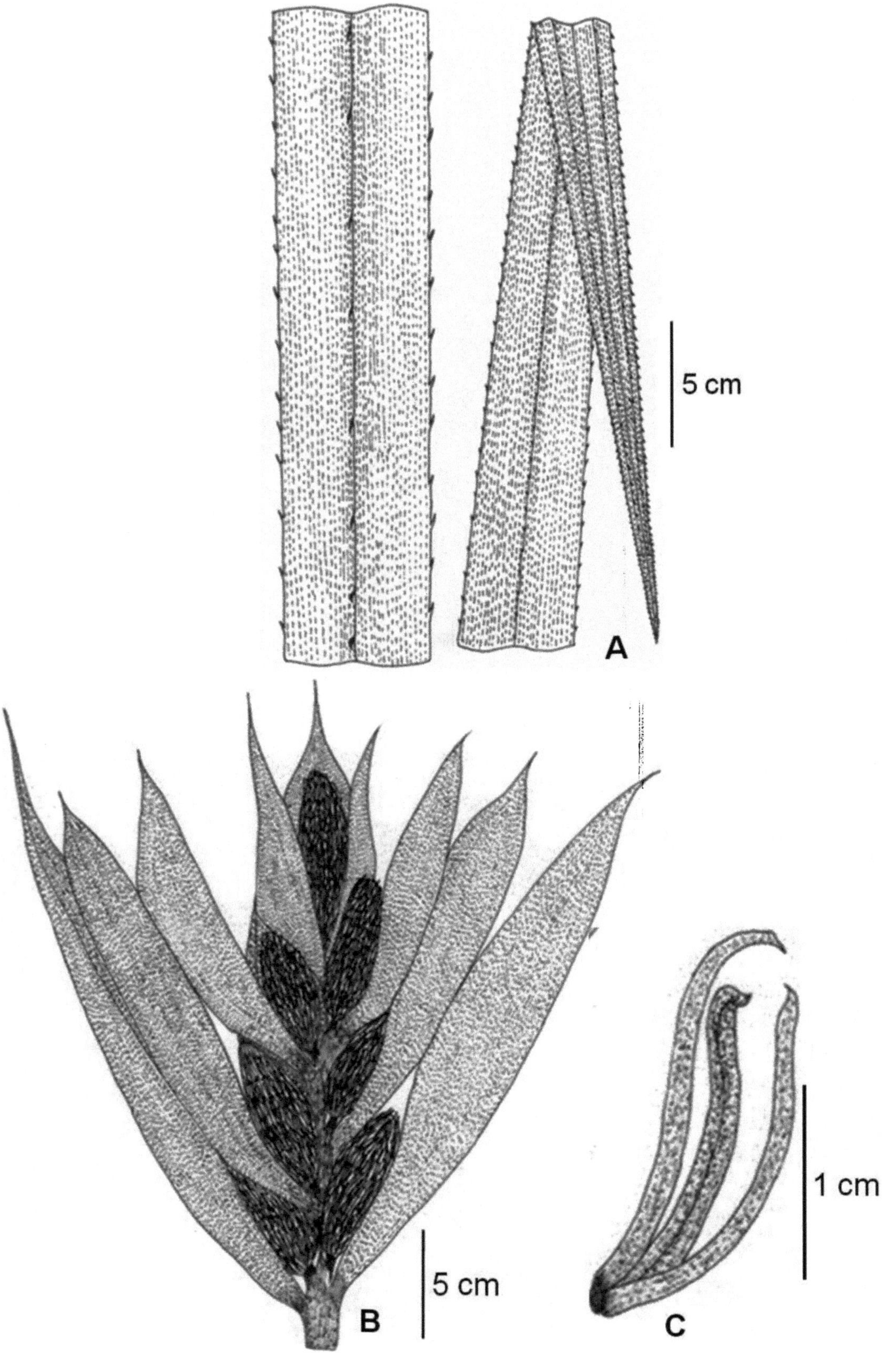

Fig. 4.49 *Benstonea thwaitesii*: (**a**) adaxial and abaxial leaf surface; (**b**) staminate inflorescence; (**c**) androphore

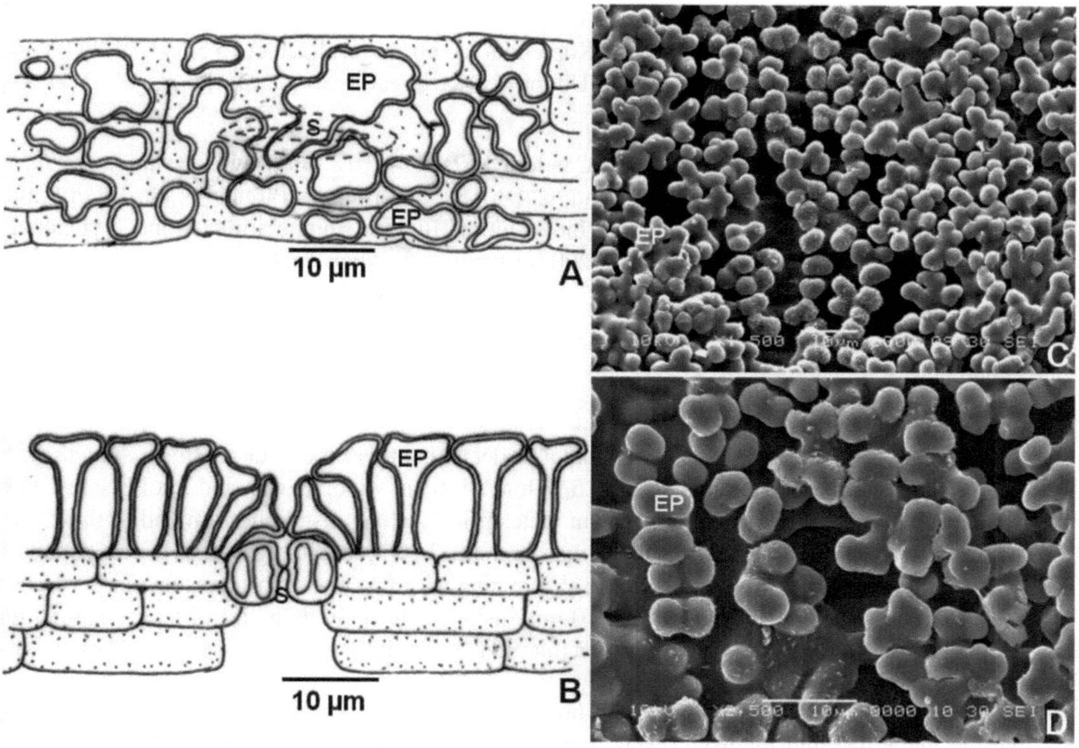

Fig. 4.50 Abaxial leaf epidermal papillae in *B. thwaitesii*: (**a**) surface view; (**b**) longitudinal section of abaxial epidermis; (**c**) SEM of papillae (×1,500); (**d**) close view of papillae (×2,500) (*S* stomata, *EP* epidermal papillae)

Flowering: December to February.

Leaf micro-morphology: (Fig. 4.50)
In *B. thwaitesii*, the abaxial epidermis is covered by papillae. The epidermal and stomatal papillae are not distinguishable; numerously branched (2–10 branches per papilla) and unbranched, supported by an individual stake, extending perpendicular to epidermis (12–18 µm long) and canopy spreading horizontal to epidermis forming thick velvety mat on the whole epidermis. According to Tomlinson (1965) and Kam (1971), *B. thwaitesii* showed Class 5 (overarching papillae lobed or dendritic) type of stomata.

Notes:
From South India, *B. thwaitesii* is observed in dense forests and sacred groves from Southern India. *B. thwaitesii* were observed near continuous natural water sources available throughout year and populations along water sources, forming dense strands. All localities represented only male populations. Inflorescence is very fragrant (similar to *P. odorifer*) and ephemeral (Zanan and Nadaf 2011). In sacred groves, male inflorescence are offered by the local people to their god.

Specimens examined:
INDIA, Maharashtra, Sindhudurg district, Phonda Ghat, Phonda sacred grove, alt. 274 m, 16°22′05″ N, 73°49′38″ E, *Rahul Zanan* 27 (staminate); Sindhudurg district, Sawantwadi, 4–5 km from Sawantwadi towards Kudal, alt. 109 m, 15°55′17″ N, 73°48′55″ E, Rahul Zanan 54 (staminate); Sindhudurg district, Danoli, 2 km from Danoli towards Sawantwadi, alt. 73 m, 15°55′51″ N, 73°55′03″ E, Rahul Zanan 55 (staminate); Sindhudurg district, Hiranyakeshi, 6–8 km from Amboli, alt. 838 m,15°57′18″ N, 74°01′22″ E, Rahul Zanan 56 (staminate); Maharashtra, Kolhapur District, Keloshi

(Radhanagari), Keloshi sacred grove, alt. 656 m, 16°30′43″ N, 73°59′05″ E, Rahul Zanan 57 (staminate); Goa, North Goa District, Valpoi, Kopardem sacred grove, alt. 26 m, 15°32′14″ N, 74°08′34″ E, Rahul Zanan 58 (staminate). Karnataka, Udupi district, Agumbe road, alt. 26 m, 13°21′54″ N, 74°52′48″ E, Rahul Zanan 29 (staminate).

Benstonea foetida (Roxb.) Callm. & Buerki Candollea 67: 333. 2012. Basionym: *Pandanus foetidus* Roxb., Fl. Ind. 3:742. 1832; Hook. F., Brit. Ind. 6:483, 1893; Warburg in Pfreich. 3: 80, 1900; Prain, BP 1101 & Veg. 294; Karthikyan et al., Fl. India, Florae Indicae Enumeration: Monocotyledonae, 177–178, 1989; Bhat in Fl. Udupi, 670–674, 2003; Bhat in Ind. J. For. 15:4, 359–360, 1992.

Original diagnosis:
Partial racemes, or *thyrses* of the male flower simple. *Germs* distinct. *Drupes* spinous pointed, with one-celled nuts.

Expanded diagnosis:
A large branched shrub about 3 m tall; stem erect or decumbent-ascending or prostrate, smooth with few prop roots at the base; Leaves up to 3–3.5 m long, 4–6 cm wide, margin and midrib spiny with three rows throughout the leaf with upper twin plates prickly along distal one-fifth end of the blade, yellow color; at the base margin with sharp spines and midrib with curved spines; at the basel part of leaves 3–4 spines on the midrib (per 10 cm), 2–3 mm long with 20–30 mm distance, on margin 10–12 spines (per 10 cm), 2.5–3 mm long with 8–10 mm distance; at the middle part of leaves 2.5–3 mm long with 7–8 spines (per 10 cm) on midrib, on margin 2.5–3 mm long with 15–18 spines (per 10 cm) and crowded towards apex.

Inflorescence terminal with yellow bracts, highly offensive odor, spinulose-denticulate on the apical margins and killed midrib; distal bracts acute, all navicular; spike sessile with 6–10 cm long; stamens crowded, directly attached to the axis, not on stemonophore, numerous, 1.5–2 cm long with very short filaments, 0.1–0.3 cm long, anthers elongated, apiculate, 1.8–2 cm long.

Infructescence terminal, highly offensive odor, on an erect peduncle; bracts with variable length, linear lanceolate, margin and midrib with prickles; carpel simple, with a linear stigma on the dorsal side of the spiniform style. Syncarps 1–3, globose up to 8 cm long, reddish on ripening; drupe oblanceolate, 5–6 angled, 2.5–3.5 cm long including the style, pileus conic, developed into a spiniform style, 0.5–0.7 cm long; linear stigma on dorsal side of the style. Endocarp abonic, 0.6–0.8 cm long; mesocarp fibrous; Seed single per locule, 0.5–0.7 × 0.3–0.4 cm (Figs. 4.51 and 4.52).

Flowering and fruiting: July to November.

Leaf micro-morphology: (Fig. 4.53)
In *B. foetida*, the abaxial epidermis is covered by papillae. The epidermal and stomatal papillae are not distinguishable, lobed, constituting 5–12 lobes per papilla, supported by single stake (6.2–8.9 μm long), tending to overarch, producing dense beautiful canopy on the stomatal chamber and epidermal cells giving totally obscuring the surface view. The papillar density increases with increase in age of plant. According to Tomlinson (1965) and Kam (1971) *B. foetida* showed Class 5 (overarching papillae lobed or dendritic) type of stomata.

Flowering and fruiting: July to November.

Notes:
Benstonea foetida is observed in dense forests at higher elevation from South India and lower elevation from Bihar, Jharkhand, and West Bengal states of India. The male and female flowers are ephemeral in nature with a highly offensive smell. During the field survey we observed that most of the inflorescences were eaten by wild animals, mostly by monkeys. *Benstonea foetida* is closely resembles *B. thwaitesii*, both the species representing genus *Benstonea* and found growing as forest undergrowth. The species differ from each other in odor, with sweet fragrant floral bracts in *B. thwaitesii* and highly offensive bracts in *B. foetida*.

Fig. 4.51 *Benstonea foetida*: (**a**) habit; (**b**) staminate inflorescence partly eaten by monkeys; (**c**) stamens; (**d**) syncarp; (**e**) stigmas in close view

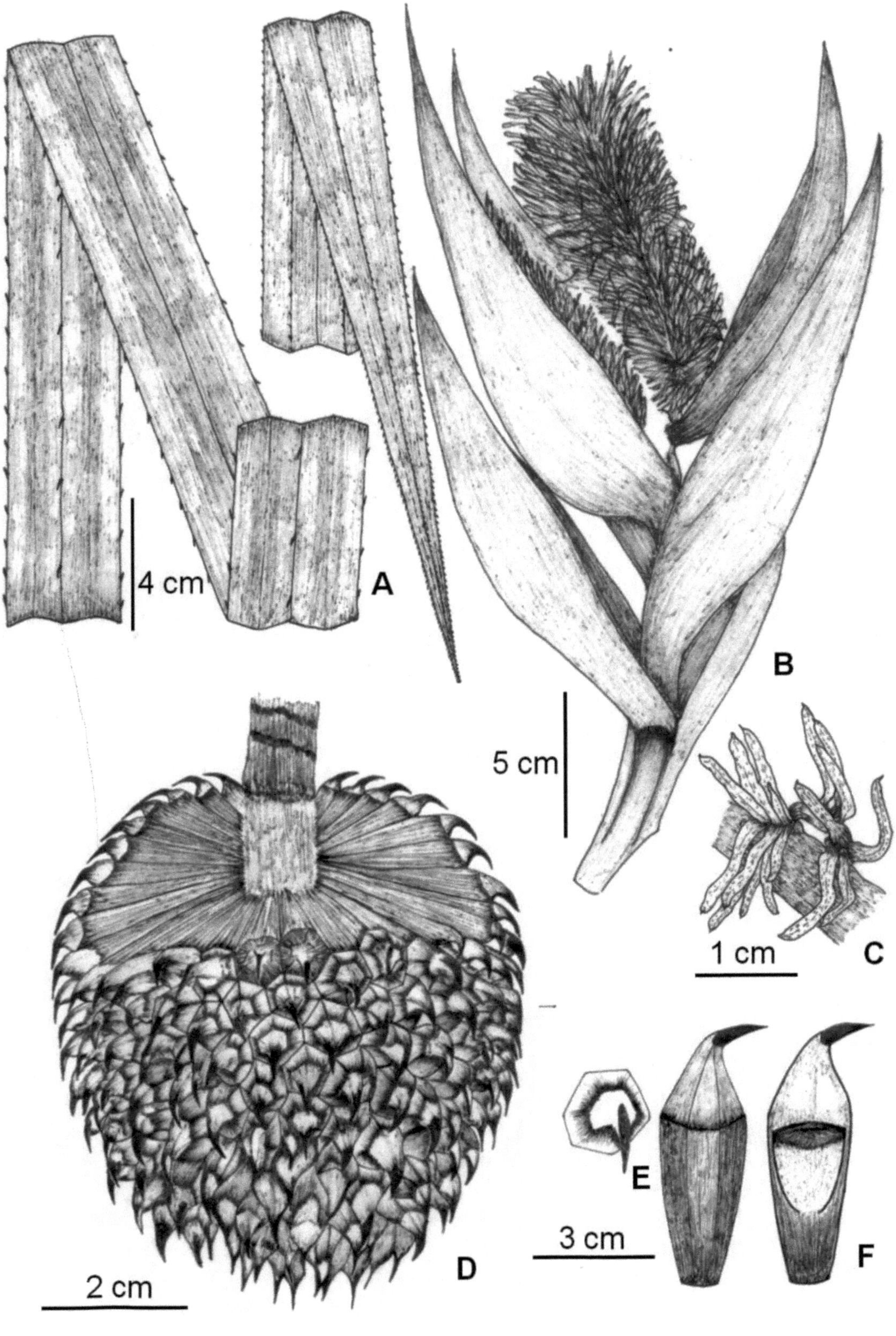

Fig. 4.52 *Benstonea foetida*: (**a**) adaxial and abaxial leaf surface; (**b**) staminate inflorescence; (**c**) androphore; (**d**) syncarp, *upper* half in longitudinal section; (**e**) stigmas in close view; (**f**) single drupe and drupe in longitudinal section showing the mesocarp and endocarp

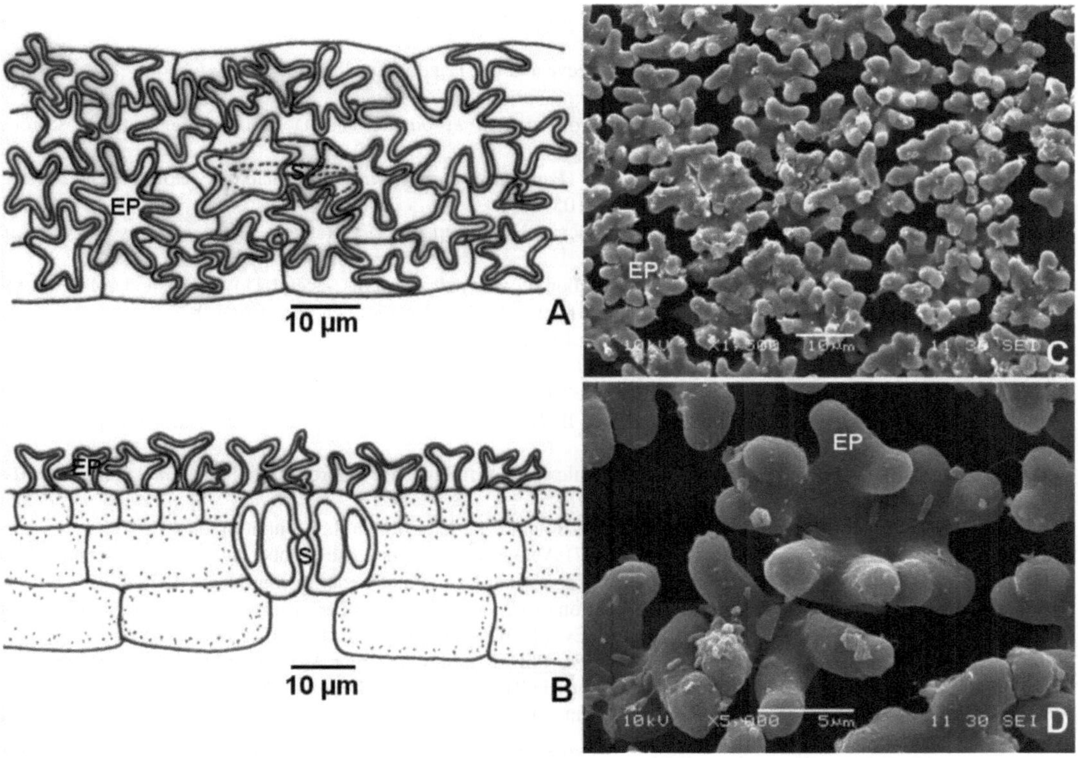

Fig. 4.53 Abaxial leaf epidermal papillae in *B. foetida*: (**a**) surface view; (**b**) longitudinal section of abaxial epidermis; (**c**) SEM of papillae (×1,500); (**d**) close view of papillae (×5,000) (*S* stomata, *EP* epidermal papillae)

Specimen examined:

INDIA, Karnataka, Shimoga district, Agumbe, alt. 170 m, 27°00′39″ N, 92°38′37″ E, *Rahul Zanan* 4 (pestilate); same locality, *Rahul Zanan* 16 (staminate); INDIA, Karnataka, Chickmangalur district, Malleshwara Forest, alt. 989, 13°27′13″ N, 95°35′31″ E *Rahul Zanan* 48 (staminate; same locality, *Rahul Zanan* 60.

References

Bhattacharjee P, Kshirsagar A, Singhal RS (2005) Supercritical carbon dioxide extraction of 2-acetyl-1-pyrroline from *Pandanus amaryllifolius* Roxb. Food Chem 91:255–259. doi:10.1016/j.foodchem.2004.01.062

Buerki S, Callmander MW, Devey DS, Chappell L, Gallaher T, Munzinger J, Haevermans T, Forest F (2012) Straightening out the screw-pines: a first step in understanding phylogenetic relationships within Pandanaceae. Taxon 61:1010–1020

Callmander MW, Chassot P, Kupfer P, Lowry PP (2003) Recognition of *Martellidendron*, a new genus of Pandanaceae, and its bio-geographic implication. Taxonomy 52:747–762

Callmander MW, Lowry II PP, Forest F, Devey DS, Beentje H, Buerki S (2012) *Benstonea* Callm. & Buerki (Pandanaceae): characterization, circumscription, and distribution of a new genus of screw-pines, with a synopsis of accepted species. Candollea 67(2):323–345

Hajra PK, Verma DM, Bandyopadhya S (1996) Flora of India, series 2, Flora of Sikkim, vol 1 (monocot.). Botanical survey of India, Calcutta

Holttum RE, St. John H (1962) Revision of the Genus *Pandanus* Stickman Harold St. John, part 11. New species from Malaya. Pac Sci 16:218–237

Hooker JD (1878) Curtis's Botanical Magazine *Pandanus unguifer* – Native of Northern Bengal, vol XXXIV, L Reeve and Co., London, p 6347
Hooker JD (1893) Flora of British India, vol VI. L. Reeve and Co, London, pp 483–488
http://www.rarepalmseeds.com/pix/PanDub.shtml
Huynh KL (1991) New data on the taxonomic position of *Pandanus eydouxia* (Pandanaceae), a species of the Mascarene Islands. Bot Helv 101:29–37
Jiang J (1999) Volatile composition of pandan leaves (*Pandanus amaryllifolius*). In: Shadidi F, Ho CT (eds) Flavour chemistry of ethnic foods. Kluwer, New York, pp 105–109
Kam YK (1971) Morphological studies in Pandanaceae III. Comparative systematic foliar anatomy of Malayan *Pandanus*. Bot J Linn Soc 64:315–351
Laohakunjit N, Noomhorm A (2004) Supercritical carbon dioxide extraction of 2-acetyl-1-pyrroline and volatile components from pandan leaves. Flavour Fragr J 19:251–259. doi:10.1016/j.foodchem.2004.01.062
Manilal KS (2003) Van Rheede's *Hortus Malabaricus*. English edition, with annotations and modern botanical nomenclature (12 vols). University of Kerala, Trivandrum
Manilal KS, Suresh CR (1984) The Pandani in Rheede's Hortus Malabaricus. New Bot 11(2):120–125
Martelli U (1913) Enumcrazione dellc 'Pandanaceae' II. *Pandanus*. Le Specie dci '*Pandanus*' enumerate per sezioni. Webbia 4(1):1–40, 94–96
Nadaf AB, Zanan RL, Wakte KV (2011) A new endemic species of Pandanaceae from India: *Pandanus palakkadensis*. Kew Bull 66:1–5. doi:10.1007/s12225-011-9265-0
Noltile HJ (1994) Flora of Bhutan, vol 3, part 1. Royal Botanic Garden Edinburgh, Edinburgh, pp 431–434
Panda KK, Panigrahy RK, Das AB, Panda BB (2007) Analysis of chromosome number, nuclear DNA content and RAPD profile in three morphotypes of *Pandanus fascicularis* Lam. Plant Genet Resour Newsl 152:12–22
Panda KK, Das AB, Panda BB (2009) Use and variation of *Pandanus tectorius* Parkinson (*P. fascicularis* Lam.) along the coastline of Orissa, India. Genet Resour Crop Evol 56:629–637. doi:10.1007/s10722-008-9390-2
Rahayu E (2011) Biosystematics of Pandanaceae in java. PhD dissertation, Bogor Agricultural University, Bogor
Solla RF (1884) Contribuzione allo studio degli stomi delle Pandanee. Nuov G bot ital 16:172–182
Srivastava RC (1997) *Pandanus unguifer* Hook. f. (Pandanaceae) a little known species from Sikkim Himalaya. Natl Acad Sci Lett 20(3 & 4):44–45
St. John H (1962) Revision of the Genus *Pandanus* Stickman, Part 10. New Pandanus Species from Vietnam. Pac Sci 16:88–125
St. John H (1963) Revision of the Genus *Pandanus* Stickman, Part 16. Species discovered in Thailand and Vietnam. Pac Sci 17:466–492
St. John H (1972) The Indian species of *Pandanus* (Section Rykia). Bot Mag Tokyo 85:241–262
St. John H (1975) Revision of the Genus *Pandanus* Stickman. Part 39 *Pandanus* of Rotuma Island, Pacific Ocean. Pac Sci 29(4):371–406
Stone BC (1974) Towards an improved infrageneric classification in *Pandanus* (Pandanaceae). Bot. Jahrb. Syst 94:459–540
Stone BC (1975) Notes on the Pandanaceae of Ceylon, with a review of the species cultivated in the Royal Botanic Gardens, Peradeniya, and those found in its Herbarium. Ceylon J Sci (Bio Sa) 11(2):109–125
Stone BC (1976) Pandanaceae. In: Saldhanha & Nicolson (eds) Flora of Hassan District Karnataka, India. Smithsonian Science Foundation/Amerind Publishing Co., Washington/New Delhi, pp 777–781
Stone BC (1978) Studies in Malesian Pandanaceae XVII on the Taxonomy of 'Pandan Wangi': a Pandanus cultivar with scented leaves. Econ Bot 32(3):285–293
Stone BC (1981) Pandanaceae. In: Dassanayake MD, Fosberg FR (eds) A revised handbook to the flora of Ceylon Vil. III. Oxford & IBH Pub. co, New Delhi, pp 293–320
Stone BC (1983) Pandanaceae. In: Matthew KM (ed) Flora of Tamil Nadu & Carnatic. The Rapinat Herbarium, Tiruchirappalli, pp 1676–1683
Sun K, DeFilipps RA (2010) Pandanaceae. In: Zhengyi W, Raven PH, Deyuan H (eds) Flora of China, vol 23. Science press/MBG press, Beijing, pp 127–130
Tomlinson PB (1965) A study of stomatal structure in Pandanaceae. Pac Sci 19(1):38–54
Van Rheede HA (1679) Hortus Malabaricus, vol 2. Sumptibus Johannis van Someren, et Joannis van Dyck, Amsterdami
Wakte KV (2010) Biochemical characterization of aroma volatiles and molecular genotyping in Indian *Pandanus amaryllifolius* Roxb. Clones. PhD dissertation, Department of Botany, University of Pune
Wakte KV, Nadaf AB, Krishnan S, Thengane RJ (2007) Studies on lower epidermal papillae, site of storage of "basmati rice" aroma compounds in *Pandanus amaryllifolius* Roxb. Curr Sci 93(2):238–242
Wakte KV, Nadaf AB, Thengane RJ, Jawali N (2009) *Pandanus amaryllifolius* Roxb. cultivated as a spice in costal regions of India. Genet Resour Crop Evol 56:735–740. doi:10.1007/s10722-009-9431-5
Warburg O (1900) Pandanaceae. In: Engler A, Das Pflanzenreich, 3(1V.9):1–99

Zanan RL, Nadaf AB (2011) Collection, characterization and olfactory evaluation of Pandanus species in Southern India. Plant Genet Resour. doi:10.1017/S1479262111000748

Zanan RL, Nadaf AB (2012a) *Pandanus mangalorensis*: a new species of Pandanaceae from Southern India. Kew Bull 67:1–5. doi:10.1007/s12225-012-9366-4

Zanan RL, Nadaf AB (2012b) Pandanus martinianus (Pandanaceae), a new endemic species from northeastern India. Phytotaxa 73:1–7

Chapter 5
Species Identification Key for Indian Pandanaceae

The Indian *Pandanus* species are classified using Stone's infrageneric classification system (1974). In Stone's classification, several sections are taxonomically difficult, with conflicting interpretation of species boundaries and their interrelationships (Stone 1974). The previously published classification systems are based on female plants only because of the ephemeral nature of staminate inflorescence, and lack of sufficient characters in leaf and diagnostic characters in fruit (Kam 1971). In the present study, we critically analyzed all Indian *Pandanus* and *Benstonea* species for the vegetative, inflorescence, and infructescence characters and *Pandanus* species are classified according to the subgenera, sections, subsections, and species level (Key 5.1). In addition, the micro-morphological characters with respect to the abaxial stomata and epidermal papillae were considered and a key is provided that could be used to identify the species at a preliminary level. The taxonomists working on Pandanaceae are of the opinion that the leaf epidermises are identical throughout the genus, with considerable difference in stomatal size and distribution pattern of papillae (Tomlinson 1965; Kam 1971; Solla 1884), which can be used for systematic purposes (Kam 1971).

Key 5.1: Key Based on the Morphological Characters

Genus: Pandanus

1a. Stamens few to numerous; carpels free or connate in phalanges, phalanges several seeded; carpel free, stigma always on the ventral side of the style, stigma variable in shape but not linear carpels connate, stigma concentrically or irregular; endocarp of carpel in phalanges completely united.

2a. Leaf apex with ventral plates always unarmed; stamens in phalanges, crowded at the apex of stemonophore; style acute, deltoid-ovate, spiniform or bifurcate, carpel free or connate into phalanges.
.... **Subg. Rykia**

3a. Leaves with unarmed apical ventral plates; style spiniform, forked, simple, or both types in one cephalium.
.... **Sect. *Rykia***

4. Stemonophore columnar; carpels free, fruits as single-seeded drupes, cephalia solitary, endocarp longer than basal mesocarp, styles nearly all or mostly forked.
.... **Subsect. *Rykia***

A. Nadaf and R. Zanan, *Indian Pandanaceae - an overview*,
DOI 10.1007/978-81-322-0753-5_5, © Springer India 2012

5a. Single cephalium per peduncle, rarely 1–3.

Prop root frequently present; large size of cephalia, style forked and acute.
.... ***P. diversus***

Prop root rarely present; cephalia triangular, pileus elevated, style forked and acute.
.... ***P. mangalorensis***

Pileus elevated, style forked, central projection is not attached to pileus base.
.... ***P. unipapillatus***

Pileus pyramidal, style forked, central projection is attached to pileus base.
.... ***P. furcatus***

Prop root rarely present at the base; cephalia cylindrical, style forked and acute, pileus plain, central projection of endocarp is attached to the pileus base with elevated shoulders.
.... ***P. palakkadensis***

5b. Multiple cephalium per peduncle.

Prop roots frequently present; 13–17 stamens; cephalia ellipsoid, 6–8 cephalia per peduncle, pileus pyramidal to conic.
.... ***P. nepalensis***

Prop roots frequently present; 12–16 stamens; cephalia ellipsoid, 4–5 cephalia per peduncle, pileus elevated, retuse at the center.
.... ***P. emerginatus***

3b. Leaves with unarmed apical ventral plates; carpels free or connate in phalanges; style unforked, broadly ovate, acute, or deltoid.
.... 6

6a. Leaf apex gradually attenuate; cephalia ellipsoid to oblong, style simple, spiniform, deltoid, carpel simple, pileus about the same length as rest of drupe.

Cephalia globular, style forked and deltoid, central projection of endocarp is attached to pileus base with distinctly elevated shoulders.
.... ***P. unguifer***

Cephalia ellipsoid, style forked and deltoid, upper two-thirds forked and lower one-third deltoid, endocarp cone shaped, central projection is attached to pileus base.
.... ***P. martinianus***

6b. Carpel free or connate in phalanges, style ovate-acute or deltoid.
.... 7

7a. Leaf epidermis clearly divided into costal and landward zones; stamens connate, umbellate to racemose; phalanges large, apical mesocarp of solid pith.
.... **Sect. *Hombronia***

Leaf apex acute, round small cephalia, pileus pyramidal, 1–2 phalanges per carpel.
.... ***P. dubius***

Leaf apex acuminate, ellipsoid small cephalia, pileus elevated, about 5–12 phalanges per carpel.
.... ***P. leram***

7b. Stamens umbellate; phalanges medium, carpels concentric.
.... **Sect. *Kaida***

Stamens umbellate; 1–3 cephalia per peduncle, 2–5 phalanges per carpel, spherical style, pileus elevated, central projection of endocarp is not attached to pileus without shoulders.
.... ***P. kaida***

2b. Stamens connate into phalanges; carpels connate into phalanges, individual carpel apices convex, separated by deep grooves, carpels irregular, stigma 'V' shaped.
.... **Subg. Pandanus**

8a. Ventral plates of leaf apex unarmed; stigma predominantly horizontal, staminal phalanges racemose; phalanges more than 20, phalange apex convex to subtruncate, carpels of each phalange arranged concentrically.
.... **Sect. *Pandanus***

Terminal inflorescence, stamens 19–23, filaments free with 0.5–2 mm long; syncarp oblong round, 25–30 cm long, 18–25 cm diameter, drupes compressed, smooth, pentagonal or hexagonal or angular, apex low rounded; each drupe contains 5–15 concentrically arranged fused carpels, phalanges free, each carpel contains single 'U'- or 'V'-shaped stigma.
.... ***P. odorifer***

8b. Stamens attached in phalanges.

.... **Subg. Kurzia**

9. Stamens in phalanges

.... **Sect. *Jenneretia***

Herbaceous, spines are at the tip only/spineless, fragrant leaves.

.... ***P. amaryllifolius***

Genus: *Benstonea*

Leaf apex with ventral plates prickly; stamens singly attached to spike axis, many crowded, anthers much longer than filaments; carpels free, stigma on the dorsal side of the style, stigma linear, single seeded drupes. Pileus smooth, style not much elongated, terrestrial plants with stilt roots, leaves linear, apical ventral plates armed, cephalia globose.

.... **Genus *Benstonea***

Leaf apex with apical ventral plates armed, inflorescence fragrant, cephalia globose with several heads (5–7).

.... ***B. thwaitesii***

Leaf apex with apical ventral plates armed, inflorescence offensive, cephalia globose with 1–3 heads.

.... ***B. foetida***

Key 5.2: Identification Key Based on Micro-morphological Characters of Indian *Pandanus* and *Benstonea* Species (Tomlinson 1965; Kam 1971)

Genus: Pandanus

1a. Papillae Present on Abaxial Epidermis

.... 2

2a. Abaxial Epidermal Papillae Restricted Around Stomatal Chamber

Epidermal papillae present on single lateral subsidiary cell or both the lateral subsidiary cells or lateral subsidiary cell with terminal subsidiary cell or stomatal chamber with adjacent epidermal cells; lateral subsidiary cell having 4–6 small, simple, globular papillae.

.... ***P. odorifer***

Papillae are only on subsidiary cells and very rarely on adjacent epidermal cells; lateral subsidiary cell having 3–5 small, simple, globular, or slightly triangular papillae.

.... ***P. nepalensis***

2b. Epidermal Papillae Around Stomata and on Whole Epidermis

.... 3

3a. Alike Distributed Epidermal Papillae

1–7 unbranched, simple and alike distributed papillae on each epidermal cell; adjacent subsidiary cell born with 3–7 unbranched papillae, producing wall around the guard cells of stomata and form beautiful necklace-like structure.

.... ***P. amaryllifolius***

Epidermal papillae attached to each other and form vertical row, opposite to cell length with 3–6 papillae; 4–5 small unbranched globular papillae on lateral subsidiary cell.

.... ***P. diversus***

3b. Randomly Distributed Epidermal Papillae

.... 4

4a. Terminal Subsidiary Cell Papillae Unbranched

Abaxial epidermal papillae long elongated unbranched dome shaped and each papilla seems to be separately born, forming rows having 3–4 similar or different size papillae; 3–5 small unbranched dome-shaped papillae on lateral subsidiary cell.

.... ***P. leram***

Epidermal papillae small, dome shaped and each papilla born separately with 1–2 papillae per cell; lateral subsidiary cell having 3–5 small unbranched low triangular papillae.

.... ***P. emerginatus***

4b. Terminal Subsidiary Cell Papillae Branched and Unlobed

Epidermal papillae small triangular, large dome shaped, unbranched and separately born without stack; 3–6 small unbranched triangular papillae on lateral subsidiary cell.
.... ***P. furcatus***

Epidermal papillae conical, dome shaped, 3–4 papillae in chain varying in size; epidermal cellular papillae taller than lateral subsidiary cell papillae; lateral subsidiary cell born with 3–5 small papillae.
.... ***P. kaida***

Epidermal papillae 1–3 dome shaped and triangular; dome-shaped papillae are bigger than triangular papillae; 3–6 smaller triangular and unbranched papillae on lateral subsidiary cell.
.... ***P. unipapillatus***

Small dome-shaped epidermal papillae; terminal subsidiary cell-borne single bifurcated, elongated or rod-shaped papilla tending to overarch guard cells of stomata; 3–5 small unbranched, low triangular to dome-shaped papillae on lateral subsidiary cell.
.... ***P. palakkadensis***

Epidermal papillae elongated, unbranched; terminal subsidiary cell-borne single long elongated and low bifurcated papilla; lateral subsidiary cell-borne 3–6 small unbranched papillae.
.... ***P. mangalorensis***

1b. Papillae Absent on Abaxial Epidermis

Large stomatal cells (25–31 μm long guard cells)
.... ***P. leram, P. martinianus, P. odorifer* and *P. dubius***

Small stomatal cells (17–19 μm long)
.... ***P. nepalensis* and *P. unguifer***

Genus: *Benstonea*

Papillae Present on Whole Abaxial Epidermis, Randomly Distributed and All Papillae Branched or Lobed

Epidermal papillae branched, rarely unbranched; constituting 2–10 branches per papilla to produce thick velvety mat on whole epidermis that gives totally obscuring surface view, most of the papillae having 2–5 branches; papillae are extended horizontally to epidermis.
.... ***B. thwaitesii***

All papillae numerously branched, long lobes constituting 5–12 lobes per papillae to produce dense canopy on whole epidermis with different heights. Density of papillae increases with lobe height, extending papillar canopy obtuse epidermal cells and stomatal chambers.
.... ***B. foetida***

References

Kam YK (1971) Morphological studies in Pandanaceae III. Comparative systematic foliar anatomy of Malayan Pandanus. Bot J Linn Soc 64:315–351

Solla RF (1884) Contribuzione allo studio degli stomi delle Pandanee. Nuov G bot ital 16:172–182

Stone BC (1974) Towards an improved infrageneric classification in *Pandanus* (Pandanaceae). Bot Jahrb Syst 94:459–540

Tomlinson PB (1965) A study of stomatal structure in Pandanaceae. Pac Sci 19(1):38–54

Chapter 6
Phylogenetic Relationship Among the Indian *Pandanus* Species

Taxonomy is a systematic classification of living organisms, whereas phylogeny is a theoretical model of the sequence of evolutionary divergence of organisms from their common ancestors. Phylogeny is derived from a combination of Greek words: *phylon* means stem and *genesis* means origin. It is the study of evolutionary relationships among organisms. Traditionally, morphology, anatomy, physiology, and paleontology are used to determine the phylogeny (Riley 2009). In Pandanaceae, the morphological characters used to describe species are mainly based on fruit. Further, characterization of the species requires a large set of phenotypic data that are difficult to access statistically and are variable due to environmental effects (Sedra et al. 1993, 1996, 1998). There are a number of DNA-based marker systems available for studying phylogeny. Unlike morphological markers, molecular markers are not prone to environmental influences and do portray the genetic relationship between plant groups (Powell 1992; Gottlieb 1977; Tanksley et al. 1989; McCouch and Tanksley 1991).

In order to illustrate phylogenetic relationships among plants, researchers have considered both morphological and molecular data (Witt 2004). In molecular phylogeny, the relationships among organisms or genes are studied by comparing homologues of DNA or protein sequences. Dissimilarities among the sequences indicate genetic divergence as a result of molecular evolution over the course of time. The molecular phylogenetic approach is increasingly useful, especially where morphological characters have been insufficient for distinguishing genera (Soltis and Soltis 1997). If the genera differ with any single character state, the phylogenetic approach allows differentiating easily. In Saxifragaceae, plant morphology has been inadequate for alignment of genera within the family because of similarities among vegetative features, with some genera distinguished by only a few pronounced differences in floral or fruit morphology (Soltis et al. 1993, 2001a). To study the phylogenetic analysis, DNA and amino acid sequences have several advantages over traditional morphological approaches, because the universality of the character types and states; the high number of characters available for analyses; the high degree of variation in the substitution rates among genes and gene regions; our increasingly comprehensive knowledge of the molecular basis underlying sequence evolution and function; and the relatively easy collection of the data from different taxa (Hillis and Wiens 2000).

Molecular phylogenies have resolved the relationships among all green plants (Källersjö et al. 1998), all major groups of extant land plants (Kranz et al. 1995; Kranz and Huss 1996; Manhart 1994; Qiu and Palmer 1998), and major groups of angiosperms (Chase et al. 1993; Soltis et al. 1997). In the past few years, molecular phylogenetics has dramatically reshaped our views of organismal relationships and evolution. This impact has been manifested at all taxonomic levels of the hierarchy of life, from the species level to kingdoms (Soltis and Soltis 2000). Presently, molecular phylogeny is used wholly in the study of evolutionary relationships. Molecular phylogenetic data provide a window into the history of speciation, extinction, dispersal, and vicariance, which led to current patterns.

The phylogeny is represented by a phylogenetic tree, which is a graphical representation of the evolutionary relationships between groups of organisms. A phylogenetic tree is composed of nodes and branches, where only one branch connects any two adjacent nodes. Nodes represent the

A. Nadaf and R. Zanan, *Indian Pandanaceae - an overview*,
DOI 10.1007/978-81-322-0753-5_6, © Springer India 2012

taxonomic units such as species, populations, individuals, or genes. Branches define the ancestral relationships where their length may be proportional to the difference between nodes. The branching pattern is known as the topology. External nodes represent extant taxonomic units and are specifically referred to as operational taxonomic units. Internal nodes represent ancestral units. A tree is additive if the distance between any two operational taxonomic units is equal to the sum of the lengths of all branches connecting them. A node is bifurcating if it has only two immediate descendant lineages, multifurcating. If more than two, a clade is defined as a group of species that have a unique common ancestor that is not shared by any other species (Riley 2009).

For construction of phylogenetic tree, molecular and morphological data is useful and necessary. These two types of data constitute independent and complementary sources of information for cross-validating hypotheses about evolutionary patterns and processes at different levels of biological organization (Mauro and Agorreta 2010). Recently, methodological advances in tree-building and divergence time estimation have greatly improved our ability to resolve the order and absolute timing of speciation events. The knowledge of gene sequences has significantly impacted angiosperm phylogeny (Hilu et al. 2003).

DNA Regions Used in Phylogenetic Analysis

Different types of molecular approaches are used for plant systematics to resolve the phylogenetic relationship; in which restriction sites' analysis, comparative sequencing, DNA rearrangements, genes, and intron loss are considered. In most phylogenetic analysis, chloroplast, nuclear, and mitochondrial genomes are used.

Nuclear DNA regions show tremendous potential for inferring phylogenies at taxonomic levels below the genus and have been used in many different organisms, from plants to bacteria (White et al. 1990; Nickrent and Soltis 1995; Soltis et al. 1997, 2001a, b). The phylogenetic analyses were worked using several nuclear gene regions, including *Phy*, *Pgi*, *adh-1*, *GapA*, 5 S rDNA spacer, ITS, 5.8 S rDNA, 26 S rDNA, and 18 S rDNA. These gene regions were used for different taxonomic levels (Nickrent and Soltis 1995).

Phylogenetic analysis using mitochondrial genome has been mainly explored in animals (Moritz et al. 1987; Harrison 1989; Avise 1986, 1994; Avise et al. 1987; Hillis et al. 1996) and it has been little used in plant phylogeny (Nickrent and Soltis 1995). The rate of nucleotide substitution in plant mitochondrial DNA is very low; it is 3–4 times slower than chloroplast DNA and 12 times slower than nuclear DNA (Palmer 1992) and these can be used at higher levels (Nickrent and Soltis 1995). The *coxI* and *atpA* genes have been successfully used in all angiosperm (Nickrent and Soltis 1995).

The chloroplast genome is highly conserved and *rbc*L, *atp*B, *mat*K, *ndh*F, *atp*B-*rbc*L intergenic spacer, *trn*L-F spacer, cpITS2 and cpITS3, rps2, and rDNA regions have been used to resolve relationships at the familial level and higher (Soltis et al. 1990, 1993, 1997, 2001a; Savolainen et al. 2000; Hilu et al. 2003; Quandt et al. 2004; Nickrent and Soltis 1995). These genes have been used in phylogeny because of their ability to be easily amplified, few insertion-deletion events, and their level of evolution and conservation (White et al. 1990; Nickrent and Soltis 1995; Hilu et al. 2003).

Chloroplast DNA Regions Used in Molecular Phylogeny

The chloroplast genome is a circular molecule characterized by two inverted repeat segments that separate into the large and small single copy regions (Nickrent and Soltis 1995; Palmer 1985a, b). The rate of chloroplast protein-coding genes is fivefold slower than nuclear genes (Wolfe et al. 1987, 1989). Because of conservative rate of nucleotide substitution cpDNA gene sequences are useful to resolve plant phylogenetic relationships at deeper level of evolution (Clegg and Zurawski 1992).

In plant phylogenetic analysis, the *rbc*L gene sequence is used to resolve the interrelationship at family or above level (Soltis and Soltis 1990; Nickrent and Soltis 1995). The *atpB* region is located downstream of *rbc*L region; these two genes are separated by an intergenic spacer, which is useful for assessing phylogenetic relationships (Nickrent and Soltis 1995). The *mat*K, a chloroplast gene is most rapidly evolving (Wolfe 1991) and *trnL-trnF* gene region is highly conserved (Michel and Dujon 1983; Cech 1988); useful for phylogenetic reconstruction (Compton et al. 1998; Bakker et al. 1999; Bayer and Starr 1999; McDade and Moody 1999; Johnson and Soltis 1994, 1995; Soltis et al. 1996, 2001a). Along with *ndh*F gene, 18 S, 5.8 S, and 26 S rDNA regions are used for phylogenetic analysis (Nickrent and Soltis 1995; Kim and Jansen 1995; Olmstead and Reeves 1995). Recently, combined analyses are used for better phylogenetic analysis than individual sequence (Pirie et al. 2007).

Statistical Methods Used in Phylogeny

Recently, many phylogenetic clustering methods have been developed (Swofford et al. 1996; Whelan et al. 2001; Felsenstein 2004; Holder and Lewis 2003), such as distance-based methods such as neighbor-joining (Saitou and Nei 1987), parsimony-based methods such as maximum parsimony (Fitch 1971; Farris 1983), character-based methods such as maximum likelihood (Hao and Qi 2003; Xiaomeng et al. 2005; Wu et al. 2005), and Bayesian inference (Richard and Olivier 2002; Reijmers et al. 1999). Parsimony is a method of inferring phylogenetic trees with the least evolutionary change (Felsenstein 1983). Maximum parsimony is one of the earliest inference methods (Fitch 1971; Farris 1983). In this method, along with sequence data, taxonomic data (character based) is also assessed phylogenetically and hence more demanding (Robertson et al. 1998).

The neighbor-joining method (NJ) is widely used for phylogenetic reconstruction (Saitou and Nei 1987; Studier and Keppler 1988), is relatively fast, and performs well when the divergence between sequences is low. The NJ method constructs trees by clustering neighboring sequences in a stepwise manner, in each step of sequence clustering and minimizing the sum of branch lengths (Saitou and Nei 1987). The maximum likelihood method allows the inference of phylogenetic trees using complex models of sequence evolution, including the ability to estimate model parameters, thus allowing simultaneous inference of patterns and processes of molecular evolution, and provides a powerful statistical framework for hypotheses testing (Mauro and Agorreta 2010). Bayesian inference is one of the recent methods (Huelsenbeck et al. 2001; Rannala and Yang 1996; Larget and Simon 1999). It is closely allied with maximum likelihood and allows complex models of sequence evolution (Mauro and Agorreta 2010). The bootstrap was introduced by Efron (1979) to obtain estimates of error in nonstandard situations by resampling the data set. Felsenstein (1985) proposed bootstrapping as a method for obtaining confidence limits on phylogenies. It has been widely used and provides assessments of "confidence" for each clade of an observed tree, based on the proportion of bootstrap trees showing that same clade.

Phylogeny of Indian Pandanaceae

Indian Pandanaceae represents three genera viz., *Pandanus, Benstonea*, and *Freycinetia*. Amongst them, genus *Pandanus* is distributed from Northeastern to Southern India and in Andaman and Nicobar Islands; genus *Benstonea* from Northeastern to Southern India; *Freycinetia* in Andaman and Nicobar Islands only. In the present study, genus *Pandanus* and *Benstonea* were considered for phylogenetic analysis. The genus *Pandanus* represents three subgenera, Rykia, Kurzia, and Pandanus. Even though, number wise, the representation of genus *Pandanus* and *Benstonea* is less, as discussed in Chap. 4, the species exhibit wide range of variation represented by 3 subgenera among 14 species (Tables 6.1 and 6.2). Among the species, region-specific endemic distribution and a wide range of habitats from coastal to terrestrial are observed; the habit varies from small shrub to large tree; high variation in leaf

Table 6.1 State wise distribution of Indian Pandanaceae *Pandanus* species

Species	States														
	GJ	MH	GA	KA	KL	TN	OR	JH	WB	BH	SK	AS	AR	ML	AN
P. odorifer (Forssk.) Kuntz.	√	√	√	√	√	√	√		√						√
P. kaida Kurz.			√	√	√	√									
P. unipapillatus Dennst.		√		√	√										
P. furcatus Roxb.		√		√	√										√
P. palakkadensis Nadaf, Zanan and Wakte					√										
P. mangalorensis Nadaf and Zanan				√											
P. dubius Spreng.				√											√
P. amaryllifolius Roxb.	√	√	√	√	√	√	√		√						
P. leram Jones.						√									√
P. unguifer Hook. f.									√		√				
P. nepalensis St. John.									√		√				
P. emerginatus St. John.													√		
P. diversus St. John.												√			
P. martinianus Nadaf and Zanan												√	√		
B. thwaitesii (Martelli) Callm. & Buerki		√	√	√	√	√									
B. foetida (Roxb.) Callm. & Buerki				√	√			√	√	√					

GJ Gujarat, *MH* Maharashtra, *GA* Goa, *KA* Karnataka, *KL* Kerala, *TN* Tamil Nadu, *AP* Andhra Pradesh, *OR* Orissa, *WB* West Bengal, *JH* Jharkhand, *BH* Bihar, *SK* Sikkim, *AS* Assam, *AR* Arunachal Pradesh, *ML* Meghalaya, *AN* Andaman and Nicobar

Table 6.2 Geographical details of the localities and collections

Species	Locality	Collector and collection number	Geographical information		
			Longitude (°E)	Latitude (°N)	Altitude (m)
P. furcatus (pistillate)	Kudal, MH	*Rahul Zanan 01* (Dept. Bot. UoP)	73°42′39″	16°05′92″	17
P. unipapillatus (pistillate)	Udupi, KA	*Rahul Zanan 02* (Dept. Bot. UoP)	74°52′88″	13°21′90″	26
P. kaida (pistillate)	Bhatkal, KA	*Rahul Zanan 03* (Dept. Bot. UoP)	74°30′28″	14°04′09″	10
P. mangalorensis (pistillate)	Padil, KA	*Rahul Zanan 31* (Dept. Bot. UoP)	74°53′54″	12°52′34″	7
P. dubius	Udupi, KA	*Rahul Zanan 11* (Dept. Bot. UoP)	74°45′25″	13°19′83″	12
P. amaryllifolius	Avarsa, KA	*Rahul Zanan 12* (Dept. Bot. UoP)	74°16′78″	14°43′26″	12
P. odorifer (pistillate)	Trivandrum, KL	*Rahul Zanan 09* (Dept. Bot. UoP)	76°53′39″	08°30′38″	0
P. palakkadensis (pistillate)	Palakkad, KL	*Rahul Zanan 19* (Dept. Bot. UoP)	67°36′94	10°47′95″	40
P. leram (staminate)	Yercaud, TN	*Rahul Zanan 14* (Dept. Bot. UoP)	78°13′42″	11°47′11″	1,400
P. unguifer (pistillate)	Mungpoo, WB	*Rahul Zanan 35* (Dept. Bot. UoP)	88°21′28″	26°58′15″	1,514
P. martinianus (pistillate)	Dhimaji, AS	*Rahul Zanan 18* (Dept. Bot. UoP)	94°32′52″	27°28′37″	107
P. diversus (pistillate)	Dalu, Silchar, AS	*Rahul Zanan 36* (Dept. Bot. UoP)	92°51′03″	24°55′11″	35
P. nepalensis (pistillate)	Testa, SK	*Rahul Zanan 37* (Dept. Bot. UoP)	88°33′68″	27°15′82″	716
P. emarginatus (pistillate)	Bhalupung, AR	*Rahul Zanan 38* (Dept. Bot. UoP)	92°38′62″	27°00′66″	170
B. thwaitesii (staminate)	Udupi, KA	*Rahul Zanan 29* (Dept. Bot. UoP)	74°52′80″	13°21′90″	26
B. foetida (pistillate)	Agumbe, KA	*Rahul Zanan 04* (Dept. Bot. UoP)	92°38′62″	27°00′66″	170

MH Maharashtra, *KA* Karnataka, *KL* Kerala, *TN* Tamil Nadu, *WB* West Bengal, *SK* Sikkim, *AS* Assam, *AR* Arunachal Pradesh

size, micromorphology; presence or absence of prop roots, when present, high root length variation; size, color, and fragrance variation in male spadices, stemonophoreus stamens to stamens directly attached to axes; monocarpellary to multicarpellary phalanges, single to multiple syncarp with variable size and shape. In contrast, some species are morphological similar to the extent that taxonomists took them as the synonyms of one another. This similarity and variability among the species provides the basis to assess the genetic relatedness among them.

The phylogenetic relationship among the *Pandanus* and *Benstonea* species based on morphological characters provides some insight to understand the interrelationship among the species. However, morphological similarity and apparent differences in fruit structure of Pandanaceae put limitations in thorough understanding of the interrelationship. Under this situation, organelle-based molecular phylogenetics helps to resolve the picture and to draw firm conclusions. Recently, molecular techniques have been applied in better understanding interrelationships among the genera of Pandanaceae. Based on cp DNA fragments-based cladistic analysis, Callmander et al. (2003) raised subgenus Martellidendron to genus level. Again Buerki et al. (2012) separated subgenus Acrostigma from other subgenera and then Callmendar et al. (2012) recognized it as a separate genus namely – *Benstonea*. The phylogenetic relationships among members of the Indian genus *Pandanus* has not been worked out so far. In the present phylogenetic study, the relationship among the Indian *Pandanus* and *Benstonea* species has been discussed at morphological and molecular levels. For molecular analysis, three chloroplast DNA sequence regions have been used: *trn*L UAA 5′exon – *trn*L UAA3′exon, *trn*L UAA 3′exon – *trn*F GAA and *atp*B – *rbc*L chloroplast intergenic spacers (Taberlet et al. 1991; Chiang et al. 1998). The comparative analysis following two methods and molecular analysis in the light of previously proposed infrageneric classification system is given.

Morphological Characters and Character States Used in Phylogenetic Analysis

A set of 19 morphological characters with their corresponding character states was selected (Table 6.3) on the basis of our own analysis and published literature (Henry et al. 1989; Karthikeyan et al. 1989; Keller 2001; Stone 1976; Sharma et al. 1996; Yoganarasimhan et al. 1981; St. John 1972). In addition, the outgroup species, *Freycinetia formosana*, was included (Shu 2004). The data were scored qualitatively and quantitatively. The taxonomic characters for individual species were scored as "0–3" for the presence of a particular character state and "?" for a missing character to generate a set of binary data (Table 6.4) and a data matrix was derived for phylogenetic determination.

Phylogenetic analyses for taxonomic characters were performed using PHYLIP program Package version 3.69 (Phylogeny inference package, Felsenstein, University of Washington, Seattle, USA, 2004). For phylogenetic analyses, parsimony method (MP) was used through the Pars program for discrete characters. The parsimonious data set was calculated using the program "Pars" with unweighted 100 multiples of data set. *F. formosana* was defined as an outgroup in the program "Pars". The tree was obtained from the consensus tree, which was identified by the majority-rule consensus method by running the program "Consense". The tree was visualized using the program FigTree Version 1.3.1 (Rambaut 2006). Bootstrap values (Felsenstein 1985) were estimated by running the program "Seqboot" to produce 100 bootstrapped data sets from the original data set. While defining the phylogram, some clades were reduced in order to correlate with the morphological data because some of the species were not supported to clustered analysis. After reduction of clades, only those clusters were selected in which the majority of the taxa are supported.

Table 6.3 Morphological characters and states used in phylogenetic analysis

Sr. no.	Characters	Character variants
Vegetative characters		
1.	Plant conformation	0 = herb; 1 = shrub; 2 = tree; 3 = lianas
2.	Prop roots	0 = rarely present; 1 = frequently present
3.	Stem nature	0 = prostrate; 1 = erect; 2 = climber
4.	Leaf apex	0 = acute; 1 = acuminate
5.	Leaf margin	0 = spines on whole leaf; 1 = at the tip only/spineless
6.	Leaf apical prickly twin plate	0 = absent; 1 = present
7.	Leaf length	0 = small ($\geq$50 cm), 1 = medium ($\geq$100 cm), 2 = Large ($\leq$100 cm)
8.	Abaxial epidermal papillae	0 = present; 1 = absent; 2 = both
9.	Density of abaxial epidermal papillae	0 = absent/sparse; 1 = dense; 2 = mixed
Reproductive characters		
10.	Number of stamens	0 = few, countable; 1 = many, crowded
11.	Position of stamens	0 = attached on the stamenophore; 1 = directly attached to spike
12.	Number of syncarp	0 = 1; 1 $\leq$ 5; 2 $\geq$ 10
13.	Fruit shape	0 = round; 1 = ellipsoid; 2 = cylindrical
14.	Fruit size	0 = small (~7–10 cm); 1 = medium (~25–30 cm); 2 = Large (~40–50 cm)
15.	Style pattern	0 = spherical; 1 = bilobed; 2 = obliquely pointed/acute; 3 = bilobed and acute
16.	Pileus structure	0 = plain; 1 = elevated; 2 = pyramidal
17.	No. of phalanges/carpel	0 = 1; 1 $\leq$ 2; 2 $\leq$ 5; 3 $\leq$ 15
18.	Endocarp shape	0 = endocarp without shoulder; 1 = endocarp with shoulder
19.	Central projection of endocarp	0 = attached to pileus; 1 = not attached to pileus

Table 6.4 Morphological character score used in phylogenetic analysis

	Character no.																		
Species	1	2	3	4	5	6	7	8	9	10	11	12	13	14	15	16	17	18	19
P. odorifer	2	1	1	0	0	0	2	2	2	0	0	0	1	1	0	1	3	0	1
P. kaida	2	1	1	0	0	0	2	0	1	0	0	1	1	1	0	1	2	0	1
P. unipapillatus	2	0	1	0	0	0	2	0	1	0	0	0	1	1	1	1	0	0	1
P. furcatus	2	0	1	0	0	0	2	0	1	0	0	0	1	1	1	2	0	0	0
P. dubius	2	1	1	1	0	0	2	1	0	0	0	0	0	1	0	2	1	?	1
P. palakkadensis	2	0	1	0	0	0	2	0	1	0	0	1	2	2	3	0	0	1	0
P. amaryllifolius	0	1	1	0	1	0	0	0	1	0	0	?	?	?	?	?	?	?	?
P. diversus	2	1	1	0	0	0	2	0	1	?	?	0	1	2	3	2	0	0	0
P. emarginatus	2	1	1	0	0	0	2	0	1	0	0	2	1	0	1	1	0	0	0
P. mangalorensis	2	0	1	0	0	0	2	0	1	0	0	0	1	0	3	1	0	0	0
P. nepalensis	2	1	1	0	0	0	2	2	2	0	0	2	1	1	1	2	0	0	0
P. martinianus	1	0	0	1	0	0	1	1	0	?	?	0	1	0	3	2	0	1	0
P. unguifer	1	0	0	1	0	0	1	1	0	0	0	0	0	0	3	2	0	1	0
P. leram	2	1	1	0	0	0	2	2	2	0	0	0	1	2	0	1	3	0	1
F. formosana	3	1	2	0	0	0	1	?	?	?	?	?	2	0	?	?	?	?	?
B. thwaitesii	1	0	0	0	0	1	2	0	1	1	1	2	0	0	2	2	0	?	?
B. foetida	1	0	0	0	0	1	2	0	1	1	1	1	0	0	2	2	0	0	1

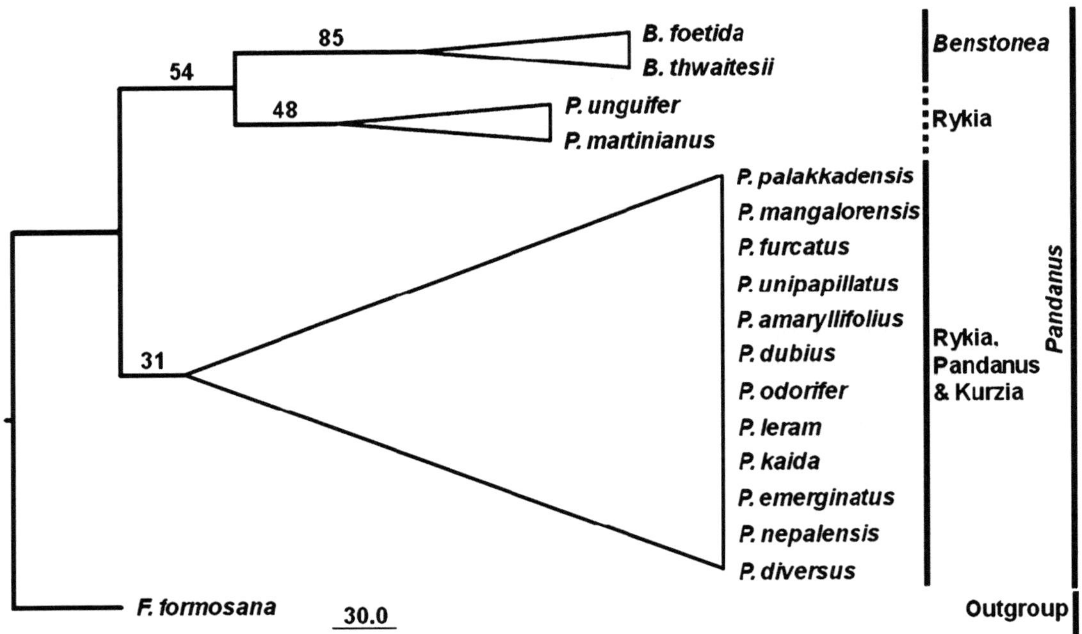

Fig. 6.1 Strict consensus morphological characters based parsimonious tree. Numbers above branch lengths indicate bootstrap support

Phylogenetic Relationship Among Pandanus Species Based on Morphological Characters

The consensus tree generated following morphological characters is shown in Fig. 6.1. All Indian *Pandanus* species were grouped in a single cluster (100% bootstrap support), keeping out the outgroup genus. The selected parsimony-informative 19 characters divided the genus *Pandanus* and *Benstonea* into three distinct clusters. *B. foetida* and *B. thwaitesii* representing genus *Benstonea* clustered together, *P. unguifer* and *P. martinianus* from subgenus Rykia formed a separate cluster and remaining species belonging to subgenus Rykia, Pandanus and Kurzia formed a single cluster.

The first cluster included two species (*B. thwaitesii* and *B. foetida*) belonging to *Benstonea* (93% bootstrapping). Genus *Benstonea* is a morphologically distinct genus sharing four distinct morphological characters of *Pandanus* species, including leaves with apical twin plate, numerous stamens directly attached to spike, and acute style.

The second cluster included *P. unguifer* and *P. martinianus* (76.5% support) representing also Subg. Rykia. These are distinct species with small-size leaves and shrubby habit. These species also shared three distinct characters in common, that is, abaxial leaf epidermis devoid of papillae, acuminate leaf apex, and endocarp with shoulders. These two species may represent a distinct group within subg. Rykia.

The third cluster is composed of 12 *Pandanus* species having erect stem, representing subgenera Rykia (*P. kaida, P. unipapillatus, P. furcatus, P. dubius, P. palakkadensis, P. diversus, P. emerginatus, P. mangalorensis, P. nepalensis* and *P. leram*), subgenus Pandanus (*P. odorifer*), and subgenus Kurzia (*P. amaryllifolius*).

The phylogram clustered the species irrespective of their geographical distribution. Further, it confirmed the distinctiveness of genus *Pandanus* from other genera. Among the genus *Pandanus*, subgenus Rykia, two species clustered separately from other species showing their close relationship both morphologically and molecularly.

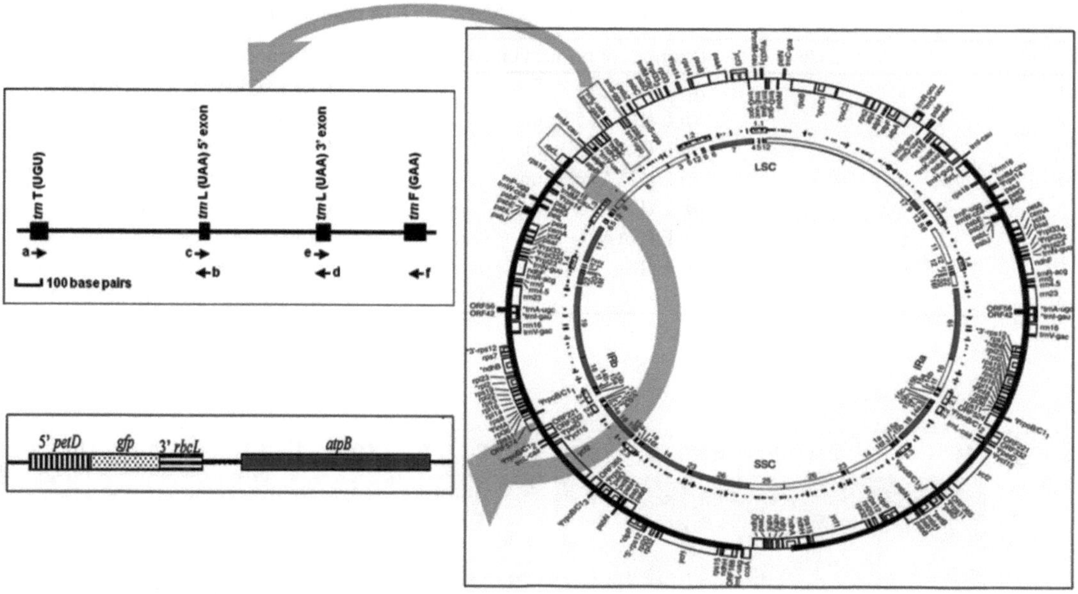

Fig. 6.2 Position and details of four cpDNA primers on cpDNA (Taberlet et al. 1991; Chiang et al. 1998)

The analysis confirms relationship among genus *Pandanus* and *Benstonea*. It further revealed the close relationship between the subgenera Rykia (Section Rykia) and Kurzia.

Chloroplast DNA (cpDNA)-Based Phylogenetic Analysis of Indian Pandanaceae

Selection of Primers

For the present study, eight previously reported chloroplast DNA-based primers were considered (*trn*T (UGU) – *trn*L (UAA) 5′ exon; *trn*L (UAA) 5′ exon – *trn*L (UAA) 3′ exon; *trn*L (UAA) 3′ exon – *trn*F (GAA) and *atp*B-1 – *rbc*L-1) (Taberlet et al. 1991; Chiang et al. 1998) (Fig. 6.2). The details of primers are mentioned in Table 6.5. The DNA samples of the respective species were amplified using four primer pairs. After verifying amplification, three primer pairs giving specific intense bands were selected for analysis.

DNA Extraction, Amplification and Sequencing

Total genomic DNA was extracted following CTAB method (Doyle and Doyle 1987) with minor modifications. The quality and quantity of the isolated DNA were determined by agarose gel (0.8%) using λ-DNA as standard (Fermentas, USA). Ethidium bromide was used as a visualizing agent and the gels were documented using gel documentation system (Bio-Rad, California, USA). The chloroplast

Table 6.5 Universal primers sequences used for the amplification of cpDNA of Indian *Pandanaceae*

Sr. no.	Primer code	Positions of primers on cpDNA	Sequence (5′–3′)
1.	a	*trn* T (UGU)	CATTACAAATGCGATGCTCT
2.	b	*trn* L (UAA) 5′ exon	TCTACCGATTTCGCCATATC
3.	c	*trn* L (UAA) 5′ exon	CGAAATCGGTAGACGCTACG
4.	d	*trn* L (UAA) 3′ exon	GGGGATAGAGGGACTTGAAC
5.	e	*trn* L (UAA) 3′ exon	GGTTCAAGTCCCTCTATCCC
6.	f	*trn* F (GAA)	ATTTGAACTGGTGACACGAG
7.	g	*atp* B-1	ACATCKARTACKGGACCAATAA
8.	h	*rbc* L-1	AACACCAGCTTTRAATCCAA

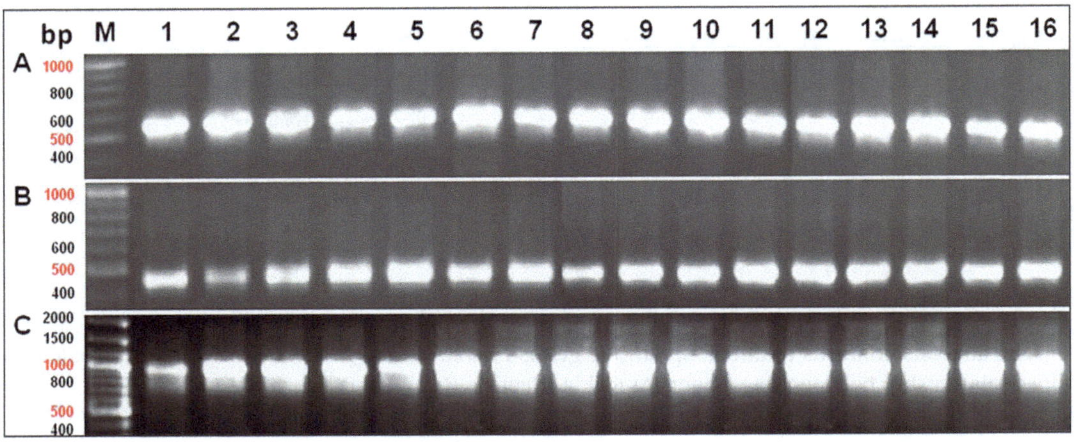

Fig. 6.3 Amplifications generated using (A) *trn*L (UAA) 5′ exon and *trn*L (UAA) 3′ exon; (B) *trn*L (UAA) 3′ exon and *trn*F (GAA) exon; (C) *atp*B and *rbc*L gene spacer primers in Indian *Pandanus* species. 1 – *P. odorifer*; 2 – *P. kaida*; 3 – *P. unipapillatus*; 4 – *P. furcatus*; 5 – *P. palakkadensis*; 6 – *P. leram*; 7 – *B. foetida*; 8 – *B. thwaitesii*; 9 – *P. dubius*; 10 – *P. amaryllifolius*; 11 – *P. martinianus*; 12 – *P. unguifer*; 13 – *P. diversus*; 14 – *P. nepalensis*; 15 – *P. emerginatus*; 16 – *P. mangalorensis* and M- 100 bp plus DNA Ladder

regions mentioned above were amplified by polymerase chain reaction (PCR). PCR amplifications were typically prepared in 25 μl reactions using 50–60 ng double stranded total genomic DNA, 0.2 mM of each dNTP (Ferments, USA), 10X PCR buffer (Invitrogen, Brazil), 1.75 mM MgCl2 (Invitrogen, Brazil), 1.5 U Taq DNA polymerase (Invitrogen, Brazil), 10 pmol of each primer (forward and reverse) (IDT, India) and purified water by volume. The DNA amplification was conducted in a thermocycler (Corbett Research, Australia). Amplifications consisted of an initial denaturation step for 5 min at 94°C followed by 40 cycles of 1 min denaturation at 94°C, 1 min annealing at 50°C, 1.30 min extension at 72°C and final extension for 5 min at 72 °C. The amplified PCR products were separated on 1% agarose gels with standard Gene Ruler 100 bp Plus DNA Ladder (Fermentas, USA) as a molecular size marker. The resulting PCR products were visualized using ethidium bromide (10 μg/ml) and documented under the gel documentation system (Bio-Rad, USA) (Fig. 6.3). The successfully amplified products were purified using ExoSAP-IT PCR clean-up kit (USB, USA) following the manufacturer's recommendations prior to sequencing. Cycle sequencing was performed according to the dideoxy chain termination method using an ABI PRISM TM BigDye TM Terminator cycle sequencing kit (Applied Biosystems) from GeneOmbio Technologies Pvt. Ltd., Pune, India.

To ensure the reliability of these markers, multiple samples from the same taxon were sequenced to determine the sequence persistency.

Alignment and Phylogenetic Analyses

Sequence files were imported separately for each species into Chromas Lite 2.01 software and exported in FASTA format (http://www.technelysium.com.au). The complementary strands were aligned; ambiguous sites were checked manually and the alignment was improved visually. These sequences were compared with previously reported *Pandanus* species sequences in the GenBank database (http://www.ncbi.nlm.nih.gov). Multiple alignments based on consensus sequences were carried out in ClustalW2 using multiple sequence alignment. For the outgroup taxa and other genera from Pandanaceae, we used previously reported sequences in GenBank (http://www.ncbi.nlm.nih.gov). The data matrices obtained were analyzed using PHYLIP program Package version 3.69 (Phylogeny inference package, Felsenstein, University of Washington, Seattle, USA, 2004).

Three data partitions corresponding to three loci were defined and combined in a single dataset. The combined matrix was then analyzed phylogenetically with the gaps treating as a missing data ("combined without gap coding"). A distance-based neighbor-joining method was performed using the "Dnadist" program. The distance matrix of the data set was calculated based on the Kimura's method (Kimura 1983) using the program "Dnadist" with unweighted 100 multiples of data set. The phylogenetic tree was prepared using the program "Neighbor" by neighbor-joining method (Saitou and Nei 1987) to obtain rooted trees. *Cyclanthus bipartitus* was defined as the outgroup in the program "Neighbor". The tree was obtained from the consensus tree was identified by the majority-rule consensus method by running the program "Consense", and finally the tree was visualized using the program FigTree Version 1.3.1 (Rambaut 2006). Bootstrap values (Felsenstein 1985) were calculated using the program "Seqboot" to generate 100 bootstrapped data sets from the original data set.

Sequence Submission

The partial chloroplast DNA sequence of *trn*L (UAA) 5′ – *trn*L (UAA) 3′, *trn*L (UAA) 3′ – *trn*F (GAA) and *atp*B-*rbc*L gene spacers determined in this study were deposited in the GenBank database (http://www.ncbi.nlm.nih.gov/GenBank). The GenBank accession numbers are given in Table 6.6.

Sequence Variation

The *trn*L (UAA) 5′ – *trn*L (UAA) 3′, *trn*L (UAA) 3′ – *trn*F (GAA) and *atp*B-*rbc*L gene spacers sequences were individually aligned in ClustralW using the Genbank sequence of *Cyclanthus bipartitus* as a reference (Callmander et al. 2003). Multiple sequence fragments from each pair of sequencing primers were aligned for Pandanaceae taxa to ensure that both strands of DNA were present for editing. The nucleotide variation and single nucleotide polymorphisms of *trn*L (UAA) 5′ – *trn*L (UAA) 3′ intron, *trn*L (UAA) 3′ – *trn*F (GAA) spacer and *atp*B-*rbc*L gene intergenic spacer were identified. Aligned sequences were manually analyzed for single nucleotide changes, replacement, and deletion of bases. The *trn*L (UAA) 5′ – *trn*L (UAA) 3′ intron sequence variations are presented in Table 6.7. *trn*L (UAA) 3′ – *trn*F (GAA) spacer sequence variations are presented in Table 6.8. *atp*B-*rbc*L gene intergenic spacer sequence variations are presented in Table 6.9.

Table 6.6 Details of chloroplast DNA sequences deposited in NCBI GenBank database

	GenBank accession numbers		
Name of taxon	trnL 5′ – trnL 3′ intron	trnL 3′ – trnF spacer	atpB-rbcL intergenic spacer
P. amaryllifolius	JQ220392	JQ220377	JQ220362
P. unipapillatus	JQ220393	JQ220378	JQ220363
P. diversus	JQ220394	JQ220379	JQ220364
P. dubius	JQ220395	JQ220380	JQ220365
P. emarginatus	JQ220396	JQ220381	JQ220366
P. furcatus	JQ220398	JQ220383	JQ220368
P. kaida	JQ220399	JQ220384	JQ220369
P. leram	JQ220400	JQ220385	JQ220370
P. mangalorensis	JQ220401	JQ220386	JQ220371
P. nepalensis	JQ220402	JQ220387	JQ220372
P. odorifer	JQ220403	JQ220388	JQ220373
P. palakkadensis	JQ220404	JQ220389	JQ220374
P. unguifer	JQ220406	JQ220391	JQ220376
B. thwaitesii	JQ220405	JQ220390	JQ220375
B. foetida	JQ220397	JQ220382	JQ220367
S. philippinesis[a]	AY337703	AY337681	AY337637
F. cumingiana[a]	AY337699	AY337677	AY337633
C. bipartitus[a]	AY337705	AY337677	AY337639

[a]Ref: GenBank accessions (http://www.ncbi.nlm.nih.gov/GenBank)

Relationship Among the Indian *Pandanus* Species Based on cpDNA

Sequence Variation

The *trn*L (UAA) 5′ – *trn*L (UAA) 3′ intron sequences showed 116 nucleotide variations, *trn*F (GAA) spacer sequences showed 102 nucleotide variations and *atp*B-*rbc*L gene intergenic spacer sequences showed 165 nucleotide variations with reference to the sequence of outgroup taxon *C. bipartitus* (Tables 6.7, 6.8, and 6.9). Thus, total 23.2% polymorphism was recorded by these three primer pairs (Table 6.10). *trn*L-*trn*L (22.1%) and *trn*L-*trn*F (21.1%) gene regions showed relatively low levels of nucleotide variation than *atp*B-*rbc*L gene intergenic spacer sequence (25.7%) (Table 6.10). The polymorphisms were recorded in terms of single nucleotide changes (SNPs), replacement and deletion of bases. Most of the bases showed deletion followed by replacement and single nucleotide changes.

Phylogenetic Analysis

The final matrix consisted of 57 sequences, representing 19 species from 5 genera. The resulting consensus sequences were about 1,650 bp in length. For most of its length, the multiple alignments were straightforward. The consensus tree is shown in Fig. 6.4, with bootstrap values indicated above the branches. The clades on the tree are well supported for family Pandanaceae and outgroup taxa. All species from different subgenera of genus *Pandanus* were fairly represented in our analysis. Family Pandanaceae is resolved as sister clade with the *C. bipartitus* from family Cyclanthaceae.

Table 6.7 The trnL gene partial DNA sequence data matrix for 18 taxa of Pandanaceae and one outgroup, *Cyclanthus bipartitus*, consisting of 116 potentially informative polymorphisms

Taxa	Nucleotide position
	11111111111111111111111111112222222222222222223333333333333333333333333333333333334444444444455
	1111122222566788011111111112222222222333335578800134456666666890112233333333334444444444555555677789244555556622
	12378156781234795621990123456789013456789012343770129840142345678541898901234567890123456789013456812523169567890123
S.philippinesis	.............CN.-...----------..-----------T......-A..--T--GTATA.ACC...---------------......T---A--T.-.ACTCT......
F.cumingiana	.............CT.-C..............-----------T......A..G.AT--GTATA.ACC-------------------....T---.--T.-TACTC-----..
P.odorifer	.............C..-...-----------.-----------T......A....AT--GTATA.ACC....--GAATATG..........T---.--TA-TACTC-----..
P.kaida	.............C..-...-----------.-----------T......A....AT--GTATA.ACC....--GAATATG..........T---.--TA-TACTC-----..
P.dubius	---TACCT...-TTC.--..C----------.-----------T.....CA....AT--GTATA.ACC....ATGAATATG..........T---.--TA-TACTC-----..
P.amaryllifolius	........GAC...C..-...----------.-----------T.....-A....ATTTATATA.ACC....ATGAATATG........TCT---.--T.-TACTC-----..
P.unipapillatus	.............C..-...-----------.-----------T......A....ATTTATATA.ACC....---------------.--..T---.--T.-TACTC-----..
P.palakkadensis	.............C..-...-----------.-----------TT....GA....ATTTATATA.ACC....---------------.--..T---.--T.-TACTC-----..
P.furcatus	.............C..-...-----------.-----------TT....GA....ATTTATATA.ACC....---------------.--..T---.--T.-TACTC-----..
P.mangalorensis	.............C..-...-----------.-----------T......A....ATTTATATA.ACC....---------------.--..T---.--T.-TACTC-----..
P.emerginatus	.............C..-...-----------.-----------T......A....ATTTATATA.ACC....---------------......T---.--T.-TACTC-----..
P.learm	.............C..-.T.-----------.-----------TT.TTCGA.A..ATTTATATA.ACC....---------------......T---.--T.-TACTC-----..
P.diversus	.............C..-...-----------.-----------T.....GA....ATTTATATA.ACC....---------------......T---.--T.-TACTC-----..
P.nepalensis	.............C..-...-----------.-----------T......A....ATTTATATA.ACC....---------------......T---.--T.-TACTC-----GG
B.martinianus	.............C..-...-----------.-----------T.....GA....ATTTATATA.ACC....---------------......T---.--T.-TACTC-----..
P.unguifer	.............C..-...-----------.-----------T.....GA....ATTTATATA.ACC....---------------......T---.--T.-TACTC-----..
B.foetida	.............C..T...-----------AAACAAG-----AT.C....A....AT--GTATAAACC..---------------------..T---.--T.CTAC.C-----..
B.thwaitesii	.............C..-...-----------AA-----------T......A....AT--GTATAAACC..---------------------..T---.--T.-TAC.C-----..
C.bipartitus	GAAAGTTATGAACCNTT-TCAAAAACAAGGGTTTG------------ATACTT-GCT$_A$C-------GGCATATAT--------ATATATATATGAATAGCTGAG-C-TCT-GATAAA

Characters are numbered consecutively along the sequence with character one referenced to nucleotide 1 of *C. bipartitus*. A dot indicates that the same nucleotide given for *C. bipartitus* is present and a dash represents a deleted base

Table 6.8 The trnL-trnF gene partial DNA sequence data matrix for 18 taxa of Pandanaceae and one outgroup, *Cyclanthus bipartitus*, consisting of 102 potentially informative polymorphisms

Taxa	Nucleotide position
	111111111111112222222222222222222222222222222222233333333333333333333333333333333333
	223333444445555678990112245556688912244555555666667777888999999999990000111122233333344444555666677778
	747023812345348148590163910893523656967236789012370489057012345678934570789578345679024682670459235892
S.philippinesis	........---A.ATTT.....T...NGT-...G.C......-----...A.............A....................-.T....-.G.-..
F.cumingiana	.G......----.TA.TT.....T.TT....AGTG.CT.....-----...A.....---------A....................-.T.C..-......
P.odorifer	.G..T....TT-..A.TTACG.TTT.......GTG.CT.G...T-----..A.............A.....-.................-.T.CG.-GGA.G.
P.kaida	CGTTT......T..A.TTACG.TTT.......GTG.CT.G.CA------..A.............A.....-..C...C....C...-.T.CC.-GGA.GC
P.dubius	.G.........-..A.TTAC..TTT.......GTG.CT.....T-----..A.............A.....-.........A.....-.T.C..GGG....
P.amaryllifolius	.G......----..A.TTAC...T........GTG.CT.....TGAAT...A.............A....................-.T.C..-GG....
P.unipapillatus	.G........--..A.TTAC...T........GTG.CT.....TGAAT.A.A....G....A....A.................T..-.T.C..-GG....
P.palakkadensis	.G........--..A.TTAC...T........GTG.CT.....TGAAT.A.A....G....A....A.................T..-.T.C..-GG....
P.furcatus	.G........--..A.TTAC...T........GTG.CTG....TGAAT.A.A....G....A....A...............A.AA..-.T.CGA-GG....
P.mangalorensis	.G........--..A.TTAC...T........GTG.CTG....TGAAT.A.A....G....A....A...............A.AA..-.T.C..-GG....
P.emerginatus	.G........--..A.TTACG..T........GTG.CTG....TGAAT.A.A....G....A....A...............A.AA..-.T.C..-GG....
P.learm	.G.......---..A.TTAC...T........GTG.CT.....TGAAT.A.A....G.........A...............A.AA..-.T.C..-GG....
P.diversus	.G....C..---..A.TTAC...T........GTGCCT.....TGAAT.A.A....G.........A...............A.AA..-.T.C..-GG....
P.nepalensis	.G.......---..A.TTAC...T........GTG.CT.....-GAAT...A.............A...........-G.GA.GAAGTGTTC..-GGA..A
B.martinianus	.G.......---..A.TTAC...T........GTG.CT.....-GAAT...A.............A.................TG...-.T.C..-GG....
P.unguifer	.G.......---..A.TTAC...T........GTG.CT.....-GAAT..TTAG-C.C....CGCGCTCCGA.T-.GG.....T....-.T.C..-GG....
B.foetida	.G...C.-----..A.TTACG..T........GTG.CT..T..-----...A.....T.........A..C..........C........-.T.C..-.G....
B.thwaitesii	.G...C.-----..A.TTAC.T.T.......AGTG.CT.....-----...A.....T........A....................-.T.C..-.G....
C.bipartitus	ATACCTTTT---GCCCCCCTAGCCCGCGTCTGACATTAATCTT-----TGCGTTTATGTAAGATTTTATTATTAGTAACATAGATGTC-ACAATG-CAGATT

Characters are numbered consecutively along the sequence with character one referenced to nucleotide 1 of *C. bipartitus*. A dot indicates that the same nucleotide given for *C. bipartitus* is present and a dash represents a deleted base

Table 6.9 The atpB-rbcL intergeneric DNA sequence data matrix for 18 taxa of Pandanaceae and one outgroup, *Cyclanthus bipartitus*, consisting of 165 potentially informative polymorphisms.

Taxa	Nucleotide position
	11111111111111122222333333333333333333333333333344444444
	1222334445566667777889001366777778899577889022222222344444455555566668999000000001
	5690794637903678901245771608195678929045482401234567894234567456789012461690123478 2
S.philippinesis	......T......-------.C.TCTA.A.-----.....T.A.G...TT.GC.T..T---....-.---.TT.......A.T
F.cumingiana	G.G...T......-------...T..A.A.-----.....T...A....T..C.T..T---.......---.TT.......-.T
P.odorifer	......T......-------G..T..A.A.-----....TT...G....T..C.T------.----.---.TT.......-.T
P.kaida	......T......-------...T..A.A.-----....TT...G....T..C.T------.----.---.TT.......-.T
P.dubius	......T......-------...T..A.A.-----....TT...G....T..C.T------.----.---.TT.......-.T
P.amaryllifolius	......T......-------...T..A.A.-----....TT.T.GT...T..C.T------..---.---.TT.......-.T
P.unipapillatus	......T......-------...T..A.A.-----....TT...G....T..C.T..TT--.----.---.TT.......-.T
P.palakkadensis	......T......-------...T..A.A.-----....TT...G....T..C.T..TT--....-.---.TT.......-.T
P.furcatus	......T......-------...T..A.A.-----....TT...G....T..C.T..TT--.----.---.TT.......-.T
P.mangalorensis	......T......-------...T..A.A.-----..C.TT...G....T..C.T..TT--.......---.TT.......-.T
P.emerginatus	......T......-------...T..A.A.-----....TT...G....T..C.T..TTC.......---.TT.......-.T
P.learm	......T......-------...T..A.A.-----....TT...G....T..C.T..TTC-.......---.TT.......-.T
P.diversus	......T......-------...T..A.A.-----....TT...G....T..C.T..TTC-.......---CTTC......-.T
P.nepalensis	......T......-------...T..ACA.-----....TT...G....T..C.T..TTC-.......---.TT.......-.T
B.martinianus	......T......-------...T..ACA.-----C...TTC.CG....T..CGTA.TTC-.....TTTGCTT....... .T
P.unguifer	......T......-------...T..ACA.-----.G-.TT...G....T..C.T..TTC-.......---.TT.......-.T
B.foetida	.A.-GGTGGGG-A-------...T..ACAC-----...CTT...G.-----.C.T..TTTT.....TC--ATT.-------AT
B.thwaitesii	......T......-------..CT..ACAC-----....TT...G.-----.C.T..ATT.----.TC--.TT.-------.T
C.bipartitus	AGAATCGATTAACGAAATAATATCACGGCTTTTTAAAGCGGCGCCCATNGTTANTT----TTTTTC---TGATAATGTT-TC

Taxa	Nucleotide position
	44444444444444444444444444444444455555555555555555555555555555555555556666666666
	11112233334566667777888888999999990001222233444444445555666666777888899900111123333
	34691204693113790238056789012345893677456736234567890134135781670368268561346723 58
S.philippinesis	.N..-.C..G.CC.NA.G..TG.A------...N...A.G..C.---------NAN......T..G.C..GN.N.ACCGCCT
F.*cumingiana*	..AAG.C..GTCCC.AGGA.TG.A------...G...AGG..C.----------...N...C.TTAG.C..GNNN.ACCNCCT
P.odorifer	.....T..........G...-----------.......G.AA.---------...T.....T..T.A......TCAAAAT
P.kaida	.....T..........G...-----------.......G.AA.---------...T.....T..T.A......TCAAAAT
P.dubius	.....T..........G...-----------.......G.AA.---------...T.....T..T.A......TCAAAAT
P.amaryllifolius	.....T.T.....C...G..T-----------.......G.AA.---------...T.....T..T.A......TTAAAAT
P.unipapillatus	A....T..........G...-----------.......G.AA.---------...T.....T..T.A......TCAAAAT
P.palakkadensis	A....T..........G...-----------.......G.AA.---------...T.....T..T.A......TCAAAAT
P.furcatus	A....T..........G...-----------.......G.AA.---------...T.....T..T.A......TCAAAAT
P.mangalorensis	A....T.T........G...-----------.......G.AA.---------...T.....T..T.A......TCAAAAT
P.emerginatus	.....T..........G...-----------...C...GTAA.---------...T.....T..T.AT....CTCAAAAA
P.learm	.....T..........G...-----------.......GTAA.---------...T.....T..T.A......TCAAAAT
P.diversus	.....T..T....C...G.C.-----------.......GTAAC---------...TC....T..T.A......TCAAAAT
P.nepalensis	.....T.......C...G...-----------.......GTAA.---------...T.....T..T.A......TCAAAAT
B.martinianus	.....T.T.....C...G..A-----------..-....GTAA.---------...T.....T..T.A......TCAAAAT
P.unguifer	.....T..........G...-----------.......GTAA.---------...T.....T..T.A......TCAAAAT
B.foetida	.....T.T.....C...G..T-----------T......G.AA.---------...T.A..TT..TTA.A.....TCAAAAT
B.thwaitesii	.....T.......C...G...-----------.......G.AA.---------...T.AT..T..T.A......TCAAAAT
C.bipartitus	GATTCCACAAAAATAGATGTCNATTTAACTATATTATGATCCNATCGCAAAACAGGGGGCACCAG-ANATTTGAANC-----

Characters are numbered consecutively along the sequence with character one referenced to nucleotide 1 of *C. bipartitus*. A dot indicates that the same nucleotide given for *C. bipartitus* is present and a dash represents a deleted base

Table 6.10 Details of cpDNA sequences variation

Sr. no.	Primer pair	Number of bp T	P	% Polymorphism
1	*trn* L (UAA) 5′ exon – *trn* L (UAA) 3′ exon	526	116	22.1
2	*trn* L (UAA) 3′ exon – *trn* F (GAA)	483	102	21.1
3	*atp* B – *rbc* L	641	165	25.7
Total		1,650	383	23.2

T total bp, *P* polymorphic bp

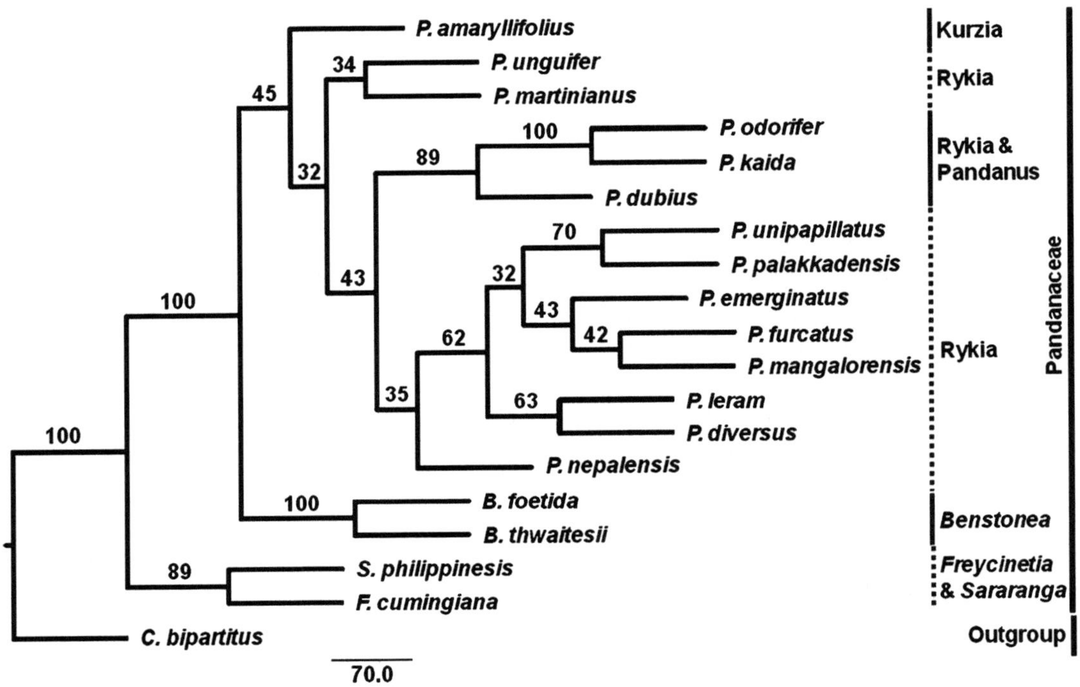

Fig. 6.4 Strict consensus cpDNA based tree of Indian Pandanaceae. Numbers above branch lengths indicate bootstrap support

One hundred percent bootstrap support was recorded for the division of the Pandanaceae into two major clades, one consisting of genus *Freycinetia* and Sararanga (89% support), and another consisting of Indian *Pandanus* and *Benstonea* species (100% bootstrap support) showed monophyletic group. The monophyletic group is represented in two distinct clusters, cluster 1 with species from subgenera Kurzia, Rykia, and Pandanus, and cluster 2 with two species from genus *Benstonea* (*B. foetida* and *B. thwaitesii*) were grouped together with 100% bootstrap support. Subgenus Kurzia showed a unique clustering pattern (cluster 1, subcluster I), was weakly supported group of single taxon *P. amaryllifolius*

(45% support). Sub-cluster 2 represents *P. unguifer* and *P. martinianus*; both the species representing the subgenus Rykia, in a distinctly grouped with low support (34%). Cluster 1, subcluster III showed polytomy, consisting of two clades, monophyletic *P. odorifer* and *P. kaida* (with 100% support) grouped with *P. dubius* (89%). *Pandanus odorifer* represented Subgenus Pandanus, Section *Pandanus*. *P. dubius* and *P. kaida* represent Subgenus Rykia, sections *Hombronia* and *Kaida,* respectively. The next branch shows large polytomous group (cluster 1, subcluster IV), include all the species from subgenus Rykia, section *Rykia* and *Hombronia* with weak bootstrap support (35%). In this cluster, *P. unipapillatus* and *P. palakkadensis* are grouped together (70%), sister to *P. emerginatus, P. furcatus,* and *P. mangalorensis*. *P. leram* and *P. diversus* are monophyletic species from sections *Hombronia* and *Rykia*, are clustered together (63% support). Within sub-cluster IV, *P. nepalensis* is grouped with all other remaining species from subgenus Rykia with low support (35%).

Sub-cluster IV belonging to subgenus Rykia appeared to be paraphyletic, although some nodes are weakly supported. In general, at the subgenus and section level, our results allow us to follow the infrageneric classification system proposed by Stone (1974) with some exception as follows. In recent studies Callmander et al. (2003) separated out the subgenus Martellidendron from genus *Pandanus* and raised it to genus level through cpDNA-based cladistic analysis. Buerki et al. (2012) separated subgenus Acrostigma from genus *Pandanus* using plastid DNA analysis; and raised to genus level as *Benstonea* (Callmander et al. 2012). In addition, based on the clustering of *P. kaida* and *P. dubius* with *P. odorifer,* we suggest shifting of *P. kaida* and *P. dubius* to Subgenus Pandanus. Stone (1974) mentioned that Martelli placed *P. kaida* under section *Hombronia* that was attributed to the Section Pandanus. In 1981, Stone shifted *P. kaida* from Section Pandanus to section Kaida (Stone 1981). Thus, its position before 1981 was in Section Pandanus and our analysis supports this placement. Our results are also supported subgenus Acrostigma as separate from genus *Pandanus* and its recognition as a separate genus *Benstonea*.

Comparison Between Molecular and Morphological Phylogenetic Analysis

The comparative account following both methods of analysis clearly reveals that the molecular analysis further refines the morphological analysis of Subgenus Rykia (Sections *Hombronia* and *Rykia*) and throws light on the interrelationships among the species. The separation of Subgenus Acrostigma from genus *Pandanus* and its upgradation to genus level (*Benstonea*) is confirmed. Moreover, the molecular analysis throws evolutionary pattern of the species under study.

Stone (1974) proposed infrageneric classification system based on morphological characters and genus *Pandanus* classified under different subgenera, sections and subsections. Morphologically, genus *Pandanus* is highly diverse genus, and there are some gaps in Stone's classification system. The family from India still lacks a strong phylogenetic understanding. In present first-time molecular phylogenetic attempts, we have evaluated Stone's classification and proposed new arrangements for the Indian genus (Table 6.11; Fig. 6.5).

Our study opens door for further analyses of the relationships within subgenus Rykia. Based upon our results, molecular markers can be developed using *atp*B-*rbc*L, *trn*L-*trn*L, and *trn*L-*trn*F sequence polymorphisms for determination of position of species under different subgenera and sections. The study proved that the chloroplast DNA-based molecular phylogenetic approach helps in better understanding the evolutionary relationship among the species.

Table 6.11 Revised classification of Indian *Pandanus* genus at subgeneric level

Genus	Group	Subgenus	Section	Subsection	Species
Pandanus Parkinson	Group A	Rykia (de Vriese) B. C. Stone	*Rykia* (de Vriese) Kurz	*Rykia* (de Vriese) Kurz	*P. furcatus* Roxb.
					P. unipapillatus Dennst.
					P. palakkadensis Nadaf, Zanan and Wakte
					P. mangalorensis Nadaf and Zanan
					P. nepalensis St. John.
					P. emarginatus St. John.
					P. diversus St. John.
					P. unguifer Hook. f.
					P. martinianus Nadaf and Zanan
			Hombronia (Gaudich.) Warb.	–	*P. leram* Jones *ex* Fontana
		Kurzia B. C. stone	*Jeanneretia* (Gaudich.) B. C. Stone	–	*P. amaryllifolius* Roxb.
	Group B	Pandanus B. C. Stone, non St. J.	*Pandanus* (syn. Keura [Forssk] Warb.)	Pandanus (syn. Keura [Forssk] Warb.)	*P. odorifer* (Forssk.) Kuntze
					P. kaida Kurz
					P. dubius Spreng.
Benstonea Callm. & Buerki					*B. thwaitesii* (Martelli) Callm. & Buerki
					B. foetida (Roxb.) Callm. & Buerki

Phylogenetic Relationship Between Indian Pandanaceae and Other Genera of Pandanaceae Outside India

It is interesting to know the position of Indian Pandanaceae in the global context. At the global level, among the various subgenera, India represents the unique Subgenus Rykia with some species from other subgenera. Therefore, an attempt has been made to assess the phylogenetic relationship between Indian Pandanus species and other genera of Pandanaceae outside India.

To determine the relationship of Indian Pandanaceae with other genera of Pandanaceae outside of India, sequences of Indian Pandanaceae members generated through our own study and reported sequences of Pandanaceae members outside India available in the GenBank database (http://www.ncbi.nlm.nih.gov) were considered along with outgroup taxa to derive the relationship. The combined data matrix was analyzed phylogenetically using PHYLIP program Package version 3.69 (Phylogeny inference package, Felsenstein, University of Washington, Seattle, USA, 2004), following the methodology mentioned elsewhere.

The final matrix represents 27 species from 5 genera. The consensus tree is shown in Fig. 6.6. In the tree, the clades well support family Pandanaceae and resolve as sister clade with the outgroup

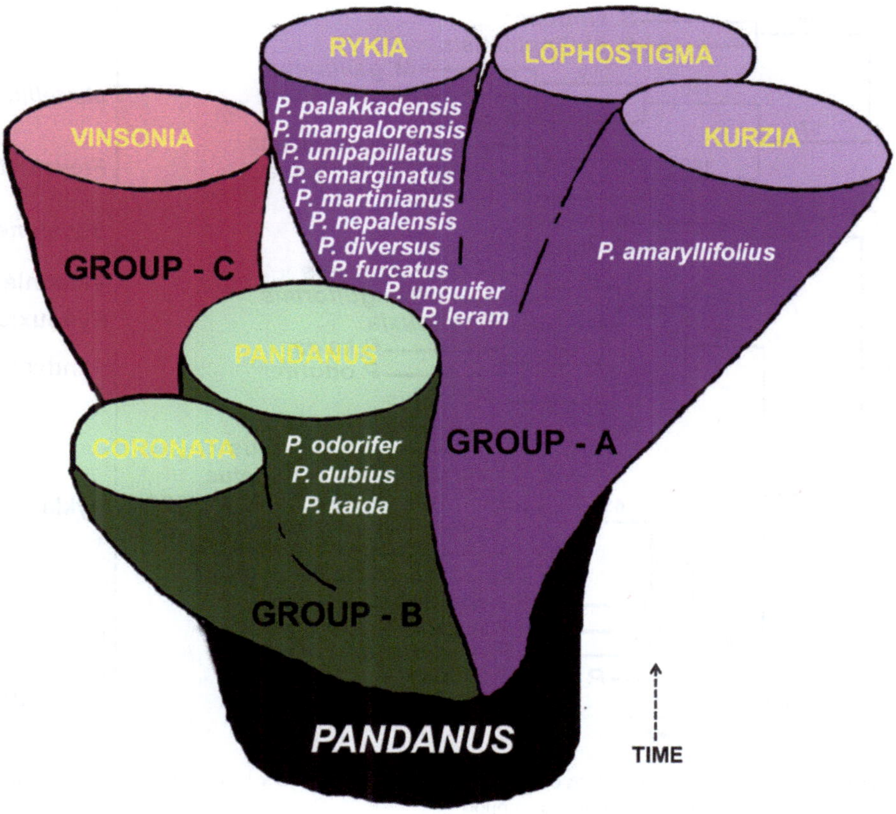

Fig. 6.5 A hypothetical infrageneric relationships in the genus *Pandanus* based on consensus evidence of the morphological characters and cpDNA sequences

Cyclanthus bipartitus Poit. ex A. Rich. (Cyclanthaceae). The family Pandanaceae is divided into two distinct clusters; in the first cluster, genus *Sararanga, Martellidendron,* and *Freycinetia* are clustered; genera *Benstonea* and *Pandanus* are clustered separately in second cluster. The phylogenetic trees indicate a close relationship between genus *Sararanga, Martellidendron,* and *Freycinetia*. A similar relationship has been reported by Callmander et al. (2003) Buerki et al. (2012) but some clades were not matched, that might be due to less no of sample size. In the second cluster, genus *Benstonea* and *Pandanus* are monophyletic, showed close relationship. Further it also confirmed the separation of subgenus Acrostigma from *Pandanus*. Within the genus *Pandanus*, the analysis clearly represents the distinction between all the subgenera considered for study. Moreover, it also confirms the shifting of *P. kaida* and *P. dubius* to Subgenus Pandanus. Subgenera Eydouxia and Vinsonia from Madagascar and Mascarene Islands revealed close relationship with Subgenus Pandanus as shown by Buerki et al. (2012). This exercise is done at primary level with the intension to understand the position of Indian screw pines in the global context. A greater number of representative species from each Subgenus needs to be explored in order to understand the interrelationship.

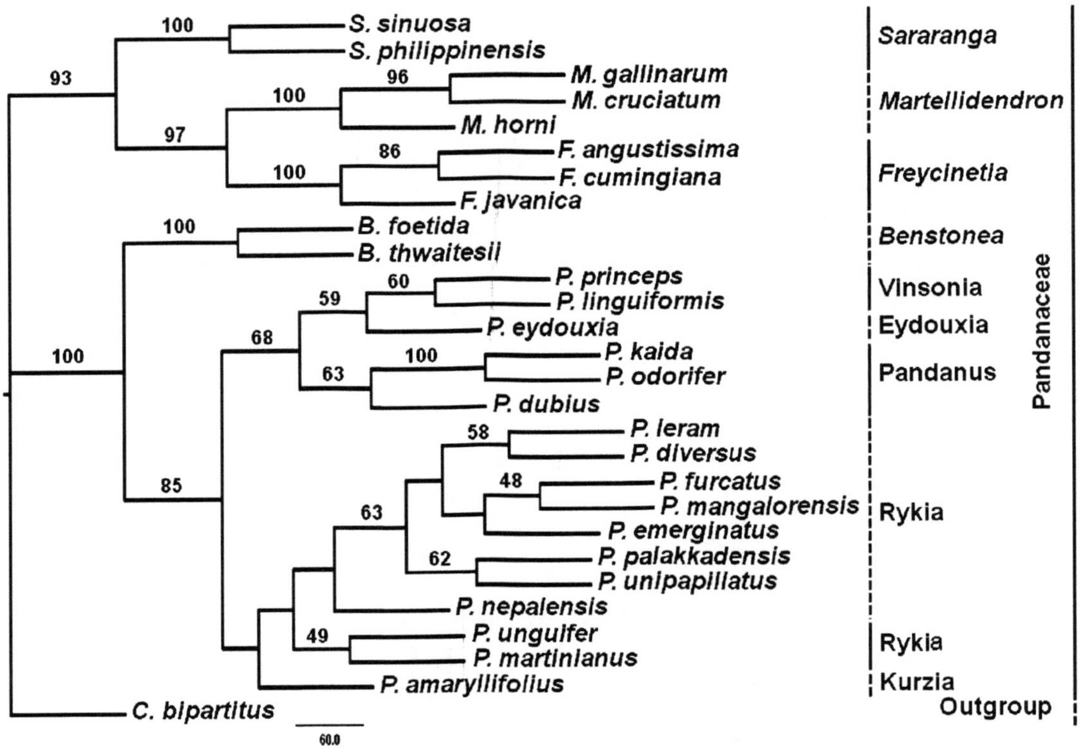

Fig. 6.6 Strict consensus cpDNA based tree of Indian Pandanaceae and other genera of Pandanaceae outside India. Numbers above branch lengths indicate bootstrap support

References

Avise JC (1986) Mitochondrial DNA and the evolutionary genetics of higher animals. Philos Trans R Soc Lond B Biol Sci 312:325–342

Avise JC (1994) The real message from Biosphere 2. Conserv Biol 8:327–329

Avise JC, Arnold J, Ball RM, Bermingham E, Lamb T, Neigel JE, Reeb CA, Saunders NC (1987) Intraspecific phylogeography: the mitochondrial DNA bridge between population genetics and systematics. Annu Rev Ecol Syst 18:489–522

Bakker FT, Culham A, Daugherty LC, Gibby M (1999) A *trnL-F* based phylogeny for species of *Pelargonium* (*Geraniaceae*) with small chromosomes. Plant Syst Evol 216:309–324

Bayer RJ, Starr JR (1999) Tribal phylogeny of the Asteraceae based on two non-coding chloroplast sequences, the *trnL* intron and *trnL/trnF* intergenic spacer. Ann Missouri Bot Gard 85:242–256

Buerki S, Callmander MW, Devey DS, Chappell L, Gallaher T, Munzinger J, Haevermans T, Forest F (2012) Straightening out the screw-pines: a first step in understanding phylogenetic relationships within Pandanaceae. Taxon 61(5):1010–1020

Callmander MW, Chassot P, Kupfer P, Lowry PP (2003) Recognition of *Martellidendron*, a new genus of Pandanaceae, and its bio-geographic implication. Taxon 52:747–762

Callmander MW, Lowry PP II, Forest F, Devey DS, Beentje H, Buerki S (2012) Benstonea Callm. & Buerki (Pandanaceae): characterization, circumscription, and distribution of a new genus of screw-pines, with a synopsis of accepted species. Candollea 67(2):323–345

Cech TR (1988) Conserved sequences and structures of group I introns: building an active site for RNA catalysis – a review. Gene 73:259–271

Chase MW, Soltis DE, Olmstead RG, Morgan D, Les DH, Mishler BD, Duval MR, Price RA, Hills HG, Qiu Y-L, Kron KA, Rettig JH, Conti E, Palmer JD, Manhart JR, Sytsma KJ, Michael HJ, Kress WJ, Karol KG, Clark WD, Hedren M, Gaut BS, Jansen RK, Kim K-J, Wimpee CF, Smith JF, Furnier GR, Strauss SH, Xiang Q-Y, Plunkett GM, Soltis PS, Swensen SM, Williams SE, Gadek PA, Quinn CJ, Eguiarte LE, Golenberg E, Learn GH Jr, Graham SW, Barrett

SCH, Dayanandan S, Albert VA (1993) Phylogenetics of seed plants: an analysis of nucleotide sequences from the plastid gene rbcL. Ann Missouri Bot Gard 80:528–580

Chiang TY, Schaal BA, Peng CI (1998) Universal primers for amplification and sequencing a noncoding spacer between the *atp*B and *rbc*L genes of chloroplast DNA. Bot Bull Acad Sin 39:245–250

Clegg MT, Zurawski G (1992) Chloroplast DNA and the study of plant phylogeny: present status and future prospects. In: Soltis PS, Soltis DE, Doyle JJ (eds) Molecular systematic of plants. Chapman and Hall, New York

Compton JA, Culham A, Jury S (1998) Reclassification of *Actaea* to include *Cimicifuga* and *Souliea* (Ranunculaceae): phylogeny inferred from morphology, nrDNA ITS and cpDNA trnL-F sequence variation. Taxon 47:593–634

Doyle J, Doyle J (1987) A rapid DNA isolation procedure for small amount of fresh leaf tissue. Phytochem Bull 5:547–555

Efron B (1979) Bootstrap methods: another look at the jackknife. Ann Statist 7:1–26

Farris JS (1983) The logical basis of phylogenetic systematics. In: Platnick NI, Funk VA (eds) Advances in cladistics. Columbia University Press, New York, pp 7–36

Felsenstein J (1983) Parsimony in systematics: biological and statistical issues. Annu Rev Ecol Syst 14:313–333

Felsenstein J (1985) Confidence limits on phylogenies: an approach using the bootstrap. Evolution 39:783–791

Felsenstein J (2004) PHYLIP (Phylogeny Inference Package) version 3.68. Department of Genetics, University of Washington, Seattle

Fitch WM (1971) Toward defining the course of evolution: minimal change for a specific tree topology. Syst Zool 20:406–416

Gottlieb LD (1977) Electrophoretic evidence and plant systematics. Ann Missouri Bot Gard 64:161–180

Hao B, Qi J (2003) Prokaryote phylogeny without sequence alignment: from avoidance signature to composition distance. In: Proceedings of the 2003 IEEE bioinformatics conference, IEEE Computer Society Washington, DC, pp 375–385

Harrison RG (1989) Animal mitochondrial DNA as a genetic marker in population and evolutionary biology. Trends Ecol Evol 4:6–11

Henry AN, Chithra V, Balakrishnan AN (1989) Flora of Tamil Nadu, India, vol III, Series 1, Analysis. Botanical Survey of India, Southern Circle, Coimbatore, 54

Hillis DM, Wiens JJ (2000) Molecules versus morphology in systematics: conflicts, artifacts, and misconceptions. In: Wiens JJ (ed) Phylogenetic analysis of morphological data. Smithsonian Institution Press, Washington, DC, pp 1–19

Hillis DM, Moritz C, Mable BK (eds) (1996) Molecular systematics. Sinauer Associates, Inc., Sunderland

Hilu KW, Borsch T, Muller K, Soltis DE, Soltis PS, Savolainen V, Chase MW, Powell MP, Alice LA, Evans R, Sauquet H, Neinhuis C, Slotta TAB, Rohwer JG, Campbell CS, Chatrou LW (2003) Angiosperm phylogeny based on *mat*K sequence information. Am J Bot 90:1758–1776

Holder M, Lewis PO (2003) Phylogeny estimation: traditional and Bayesian approaches. Nat Rev Genet 4:275–284

Huelsenbeck JP, Ronquist FR, Nielsen R, Bollback JP (2001) Bayesian inference of phylogeny and its impact on evolutionary biology. Science 294:2310–2314

Johnson LA, Soltis DE (1994) matK DNA-Sequences and phylogenetic reconstruction in Saxifragaceae S-Str. Syst Bot 19:143–156

Johnson LA, Soltis DE (1995) Phylogenetic inference in Saxifragaceae *sensu stricto* and *Gilia* (Polemoniaceae) using *mat*K sequences. Ann Missouri Bot Gard 82:149–175

Källersjö M, Farris JS, Chase MW, Bremer B, Fay MF, Humphries CJ, Petersen G, Seberg OL, Bremer K (1998) Simultaneous parsimony jackknife analysis of 2538 rbcL DNA sequences reveals support for major clades of green plants, land plants, seed plants and flowering plants. Plant Syst Evol 213:259–287

Karthikeyan SK, Jain SK, Nayar MP, Sanjappa M (1989) Flora of India, Series 4, Florae Indicae Enumeratio: Monocotyledonae. Botanical Survey of India, Calcutta, pp 177–178

Keller J (2001) Pandanaceae. In: Hanelt P, Institute of Plant Genetics and Crop Plant Research (eds) Mansfield's encyclopedia of agricultural and horticultural crops, vol 5. Springer, Berlin, pp 2816–2824

Kim KJ, Jansen RK (1995) ndhF sequence evolution and the major clades in the sunflower family. Proc Natl Acad Sci USA 92:10379–10383

Kimura M (1983) The neutral theory of molecular evolution. Cambridge University Press, Cambridge

Kranz HD, Huss VAR (1996) Molecular evolution of ferns and allies, and their relationship to seed plants: evidence from complete 18 S rRNA gene sequences. Plant Syst Evol 202:1–11

Kranz HD, Miks D, Siegler M-L, Capesius I, Sensen CHW, Huss VA (1995) The origin of land plants: phylogenetic relationships between Charophytes, Bryophytes, and vascular plants inferred from complete small subunit ribosomal RNA gene sequences. J Mol Evol 41:74–84

Larget B, Simon DL (1999) Markov chain Monte Carlo algorithms for the Bayesian analysis of phylogenetic trees. Mol Biol Evol 16:750–759

Manhart JR (1994) Phylogenetic analysis of green plant rbcL sequences. Mol Phyl Evol 3:114–127

Mauro DS, Agorreta A (2010) Molecular systematics: a synthesis of the common methods and the state of knowledge. Cell Mol Biol Lett 15:311–341

McCouch SR, Tanksley SD (1991) Development and use of restriction fragment length polymorphism in rice breeding and genetics. In: Khush GS, Toenniessen GH (eds) Rice biotechnology. CAB International and IRRI, Wallingford/Oxon/Manila, pp 109–133

McDade LA, Moody ML (1999) Phylogenetic relationships among Acanthaceae: evidence from non-coding trnL-trnF chloroplast DNA sequences. Am J Bot 86:70–80

Michel F, Dujon B (1983) Conservation of RNA secondary structures in two intron families including mitochondrial, chloroplast and nuclear-encoded members. EMBO J 2:33–38

Moritz C, Dowling TE, Brown WM (1987) Evolution of animal mitochondrial DNA relevance for population biology and systematics. Annu Rev Ecol Syst 18:269–292

National Center for Biotechnology Information. http://www.ncbi.nlm.nih.gov. Access on 26 Mar 2012

National Center for Biotechnology Information. http://www.ncbi.nlm.nih.gov/GenBank. Access on 26 Mar 2012

Nickrent DL, Soltis DE (1995) A comparison of angiosperm phylogenies from nuclear 18 s rDNA and *rbc*L sequences. Ann Missouri Bot Gard 82:208–234

Olmstead RG, Reeves PA (1995) Evidence for the polyphyly of the Scrophulariaceae based on chloroplast rbcL and ndhF sequences. Ann Missouri Bot Gard 82:176–193

Palmer JD (1985a) Comparative organization of chloroplast genomes. Annual Rev Genet 19:325–354

Palmer JD (1985b) Chloroplast DNA and molecular phylogeny. Bioessays 2:263–267

Palmer JD (1992) Mitochondrial DNA in plant systematics: applications and limitations. In: Soltis PS, Soltis DE, Doyle JJ (eds) Molecular systematics of plants. Chapman & Hall, New York, pp 36–49

Pirie MD, Vargas MPB, Botermans M, Bakker FT, Chatrou LW (2007) Ancient paralogy in the cpDNA trnL–F region in Annonaceae: implications for plant molecular systematics. Am J Bot 94:1003–1016

Powell W (1992) Plant genomes, gene markers and linkage maps. In: Moss JP (ed) Biotechnology and crop improvement in Asia. ICRISAT, Hyderabad, pp 297–322

Qiu YQ, Palmer JD (1998) Phylogeny of early land plants: insights from genes and genomes. Trends Plant Sci 4:26–30

Quandt D, Müller K, Stech M, Frahm J-P, Frey W, Hilu KW, Borsch T (2004) Molecular evolution of the chloroplast trnL-F region in land plants. Monogr Syst Bot Missouri Bot Gard 98:13–37

Rambaut A (2006) Figtree version 1.0, Institute of Evolutionary Biology, University of Edinburgh, UK. Available on http://tree.bio.ed.ac.uk/software/figtree/

Rannala B, Yang Z (1996) Probability distribution of molecular evolutionary trees: a new method of phylogenetic inference. J Mol Evol 43:304–311

Reijmers TH, Wehrens R, Daeyaert FD, Lewi PJ, Buydens LM (1999) Using genetic algorithms for the construction of phylogenetic trees: application to G-protein coupled receptor sequences. Biosystems 49:31–43

Richard D, Olivier G (2002) Fast and accurate phylogeny reconstruction algorithm based on the minimum-evolution principle. WABI2002, LNCS2452, pp 357–374

Riley MC (2009) Significant pattern discovery in gene location and phylogeny. Ph.D. dissertation, University of Wales

Robertson B, Myers G, Howard C, Brettin T, Bukh J, Gaschen B, Gojobori T, Maertens G, Mizokami M, Nainan O, Netesov S, Nishioka K, Shin-i T, Simmonds P, Smith D, Stuyver L, Weiner A (1998) Classification, nomenclature, and database development for hepatitis C virus (HCV) and related viruses: proposals for standardization. Arch Virol 143(12):2493–2503

Saitou N, Nei M (1987) The neighbor-joining method: a new method for reconstructing phylogenetic trees. Mol Biol Evol 4(4):406–425

Savolainen V, Chase MW, Hoot SB, Morton CM, Soltis DE, Bayer C, Fay MF, De Bruijn AY, Sullivan S, Qiu YL (2000) Phylogenetics of flowering plants based on combined analysis of plastid *atp*B and *rbc*L gene sequences. Syst Biol 49:306–362

Sedra MH, Filali HEL, Frira D (1993) Observations sur quelques caracteristiques phenotypiques et agronomiques du fruit des varieties et clones du palmier dattier selectionnes. Al Awamia 82:105–120

Sedra MH, Filali HEL, Benzine A, Allaoui M, Nour S, Boussak Z (1996) La palmeraie dattiere marocaine: evaluation du patrimonie phenicicole. Fruits 1:247–259

Sedra MH, Lashermes P, Trouslot P, Combes M, Hamon S (1998) identification and genetic diversity analysis of date palm (*Phoenix dactylifera* L.) varieties of Morocco using RAPD markers. Euphytica 103:75–82

Sharma DB, Karthikeyan S, Singh NP (1996) Flora of Maharashtra state, Monocotyledons. Botanical Survey of India, Calcutta, 206

Shu TLDS (2004) *Freycinetia* Gaudichaud. Ann Sci Nat (Paris) 3:509. 1824. In: Ke LDS, Kun S, DeFilipps RA (eds) Flora of China, vol XXIII. Science press, Beijing, pp 127–128

Soltis PS, Soltis DE (1990) Evolution of inbreeding and outcrossing in ferns and fern–allies. Plant Species Biol 5:1–11

Soltis DE, Soltis PS (1997) Phylogenetic relationships in Saxifragaceae *sensu lato*: a comparison of topologies based on 18 S rDNA and *rbc*L sequences. Am J Bot 84:504–522

Soltis DE, Soltis PS, Clegg MT, Durbin M (1990) *rbc*L sequence divergence and phylogenetic-relationships in Saxifragaceae *sensu-lato*. Proc Natl Acad Sci USA 87:4640–4644

Soltis DE, Morgan DR, Grable A, Soltis PS, Kuzoff R (1993) Molecular systematics of Saxifragaceae-*sensu-stricto*. Am J Bot 80:1056–1081

Soltis DE, Kuzoff RK, Conti E, Gornall R, Ferguson K (1996) *mat*K and *rbc*L gene sequence data indicate that *Saxifraga* (Saxifragaceae) is polyphyletic. Am J Bot 83:371–382

Soltis DE, Soltis PS, Nickrent DL, Johnson LA, Hahn WJ, Hoot SB, Sweere JA, Kuzoff RK, Kron KA, Chase MW, Swensen SM, Zimmer EA, Chaw SM, Gillespie LJ, Kress WJ, Sytsma KJ (1997) Angiosperm phylogeny inferred from 18 S ribosomal DNA sequences. Ann Missouri Bot Gard 84:1–49

Soltis ED, Soltis PS (2000) Contribution of plant molecular systematics to studies of molecular evolution. Plant Mol Biol 42:45–75

Soltis DE, Kuzoff RK, Mort ME, Zanis M, Fishbein M, Hufford L, Koontz J, Arroyo MK (2001a) Elucidating deep-level phylogenetic relationships in Saxifragaceae using sequences for six chloroplastic and nuclear DNA regions. Ann Missouri Bot Gard 88:669–693

Soltis DE, Tago-Nakazawa M, Xiang QY, Kawano S, Murata J, Wakabayashi M, Hibsch-Jetter C (2001b) Phylogenetic relationships and evolution in *Chrysosplenium* (Saxifragaceae) based on *mat*K sequence data. Am J Bot 88:883–893

St. John H (1972) The Indian species of *Pandanus* (section *Rykia)*. Bot Mag Tokyo 85:241–262

Stone BC (1974) Towards an improved infrageneric classification in *Pandanus* (Pandanaceae). Bot Jahrb Syst 94:459–540

Stone BC (1976) Pandanaceae. In: Saldanha CJ, Nicolson DH (eds) Flora of Hassan district Karnataka, India. Amerind Pub. Co, Calcutta, pp 777–781

Stone BC (1981) Pandanaceae. In: Dassanayake MD, Fosberg FR (eds) A revised hand book to the flora of cylon, vol III. Oxford and IBH Pub. Co., New Delhi, pp 293–320

Studier JA, Keppler KJ (1988) A note on the neighbor-joining algorithm of Saitou and Nei. Mol Biol Evol 5:729–731

Swofford DL, Olse GJ, Waddell PJ, Hillis DM (1996) Phylogenetic inference. In: Hillis DM, Moritz C, Mable BK (eds) Molecular systematics. Sinnauer Associates, Sunderland, pp 407–514

Taberlet P, Gielly L, Patou G, Bouvet J (1991) Universal primers for amplification of three non-coding regions of chloroplast DNA. Plant Mol Biol 17:1105–1109

Tanksley SD, Young ND, Paterson AH, Bonierbale MW (1989) RFLP mapping in plant breeding: new tools for an old science. Bio/Technology 7:257–264

Whelan S, Liò P, Goldman N (2001) Molecular phylogenetics: state-of-the-art methods for looking into the past. Trends Genet 17:262–272

White T, Bruns T, Lee S, Taylor J (1990) Amplification and direct sequencing of fungal ribosomal RNA genes for phylogenetics. In: Innis M, Gelfand D, Sninsky J, White T (eds) PCR protocols: a guide to methods and applications. Academic, San Diego, pp 315–322

Witt CC (2004) Rates of molecular evolution and their application to neotropical avian biogeography. Ph.D. dissertation, Louisiana State University

Wolfe KH (1991) Protein coding gene in chloroplast DNA: compilation of nucleotide sequences, data base entries and rates of molecular evolution. In: Vasil K (ed) Cell culture and somatic cell genetics of plants, vol 7BI. Academic, San Diego, pp 467–482

Wolfe KH, Li WH, Sharp PM (1987) Rates of nucleotide substation vary greatly among plant mitochondrial, chloroplast and nuclear DNAs. Proc Natl Acad Sci USA 84:9054–9058

Wolfe KH, Sharp PM, Li WH (1989) Rates of synonymous substitution in plant nuclear genes. J Mol Evol 29:208–211

Wu X, Wan XF, Xu D, Lin GH (2005) Whole genome phylogeny based on clustered signature string composition. In: Posters in 2005 IEEE computational systems bioinformatics conference (CSB2005), Stanford University, pp 53–54

Xiaomeng W, Xiufeng W, Gang W, Dong X, Guohui L (2005) Phylogenetic analysis using complete signature information of whole genomes and clustered Neighbor-Joining method. Int J Bioinform Res Appl 2:219–248

Yoganarasimhan SN, Subramanyam K, Razi BA (1981) Flora of Chikmagalur district, Karnataka. International Book Distributors, Dehra Dun

Chapter 7
Economical Importance of Indian *Pandanus* Species

Genus *Pandanus* is a large shrub or small tree of immense cultural, health, and economic importance in the Pacific (Mohsin et al. 2008). The genus is a highly variable species complex; it grows wild, mainly in seminatural vegetation, where it can withstand drought, strong winds, and salt spray (Thomson et al. 2006). Among the species only 1% has been chemically investigated (Takayama et al. 2000). It grows fairly quickly and all parts are used, from the nutritious fruits of edible varieties, to the poles and branches in construction, to the leaves for weaving and garlands (Thomson et al. 2006). A wide variety of studies have been performed in Pandanaceae family: conservation proposals (St. John 1963); phytochemical, chemosynthetic, bioinformatics, biogenetic work (Callmander 2001; Callmander et al. 2003); taxonomic research (Stone 1983); mycological (McKenzie et al. 2001), zoological (Hunt and Gray 2004; Clayton 2007), and entomological observation (Samarawickrema et al. 2008); and ethnobotany and social studies (Novellino and Ertuğ 2005). In the future, Pandanaceae will gain more attention in pharmaceutical investigations, food research, and urban landscape improvement (Mohsin et al. 2008).

The Pandanaceae members in general and Indian Pandanaceae in particular are characterized by having good fiber content, fragrant spadix, and edible fruits in some species. A detailed survey along the Malabar region and Northeastern Himalayan region of India revealed that, among the species, only *P. odorifer* and *P. amaryllifolius* are commercially exploited (Zanan and Nadaf 2011). *P. amaryllifolius* has fragrant leaves with an aroma resembling Basmati rice flavor and, for this reason, it is cultivated along the coastal parts of India (Wakte et al. 2009). The staminate spadices of *Pandanus* species are fragrant and hence, among the 16 species, *P. odorifer* has been commercially exploited for the extraction of essential oils (Panda et al. 2009). During the field survey and literature review, it was observed that, besides the *P. amaryllifolius* and *P. odorifer*; other *Pandanus* species also have own importance for their fragrance and/or for pharmaceutical and food research because of its active compounds, but, unfortunately, these species have remain largely ignored by the scientific world (Zanan and Nadaf 2011).

Pandanus odorifer

P. odorifer is a dominant coastal species in South and Southeast Asia, and is found effective in providing protection from tsunami damage due to its density and complex aerial root structure (Tanaka et al. 2007). The whole plant is used as wind breaks and protection from tsunami (Tanaka et al. 2007, 2009, 2011; Thuy et al. 2009). It is also used as a hedge plant and in preparation of pots, drums, and wood products, and in Kerala, live plants are used to prevent entry of wild animals in fields (Wilkinson and Elevitch 2000; Prasad and Raveendran 2010; Zanan and Nadaf 2011).

A. Nadaf and R. Zanan, *Indian Pandanaceae - an overview*,
DOI 10.1007/978-81-322-0753-5_7, © Springer India 2012

Pandanus odorifer and its close morphological relative from Southeast Asia (Philippines and Indonesia) and the Pacific region, *Pandanus tectorius*, are often cultivated and widely but also widely used in their native range (see Thomson et al. 2006 for more details). The trunk and large branches are commonly used as building materials in house construction. Trunks yield hard pillows, vases, and fish traps; they are also used in preparation of glue to extract cream from grated coconuts and in string making. Branches are used as wood fuel and to make compost (Thomson et al. 2006). The wood is also used for water piping (Eland 2012). Prop or aerial roots are used in fabrication of house walls, as supports, basket handles, paintbrushes, skipping ropes, handicrafts, garland string, and for dyes (Thomson et al. 2006). In some of the Pacific atolls, the roots of *P. tectorius* have been used to make a black dye and seeds have been used as stain for barkcloth (Wilkinson and Elevitch 2000; Eland 2012).

In Sri Lanka, fresh leaves of *P. odorifer* are often used for cooking and the inner parts of leaves are eaten as a vegetable (Keller 2001; Takeda et al. 2008). The tender white bases of the leaves are eaten raw or boiled (Eland 2012). In Southeastern Asia, leaf buds of *P. tectorius* used as a vegetable, and dry leaves are used for thatching, compost, matting, fans, baskets, handicrafts, jewelry, thatch, food wrapping, roofing, pillows, toys, plaited wares, cigarette wrappers, balls for children's games, ornaments, and matting sails; the fiber has been used for cordage and basketry, umbrellas, paper, wickerwork, fans, brushes for white-washing, and painting; young leaves are plaited into hats (Wilkinson and Elevitch 2000; Keller 2001; Thomson et al. 2006; Eland 2012). The young leaves are also used for lancing boils, making fans, decoration, and pig feed (Thomson et al. 2006).

In parts of India and Sri Lanka, flower water is used to flavor sugar syrups, confectionery, and soft drinks (Eland 2012). Staminate flowers are offered to gods and used for decoration (Wilkinson and Elevitch 2000; Zanan and Nadaf 2011). In Myanmar, fragrant male flower spikes have been worn in the hair, and spikes are placed among clothing for fragrance (Eland 2012). The distilled flower oil is used as an ingredient for perfumes, especially in India, where they are popular among Hindu women (Eland 2012). In India, *P. odorifer* is found in the whole South Peninsular and is abundantly distributed throughout the east and west coast of India; the plant has been valued as a key natural bioresource, particularly for perfume, at the coastal district of Ganjam in Orissa (Dutta et al. 1987; Panda et al. 2001, 2007; Sahu and Misra 2007). In the Ganjam district of Orissa, it is widely cultivated (Panda et al. 2007, 2009; Zanan and Nadaf 2011). It is estimated that about 35 million flowers (~3,500 t) are processed annually to produce fragrance and flavor materials worth Rs. 400 million (US$ 8.9 million) (Anonymous 1996). The peculiar sweet fragrant smell of *P. odorifer* flowers represents 85 volatile oil compounds, mainly due to the major constituents: 2-phenyl ethyl methyl ether (37.7%), terpinen-4-ol (18.6%), *a*-terpineol (8.3%), and 2-phenyl ethyl alcohol (7.5%) (Raina et al. 2004). In India, perfume is added to cosmetics, soaps, hair oils, and lotions, as well as being used to scent clothing and provide additional fragrance for incense and flavoring in tobacco (Eland 2012).

Fruits are used as a food stuffs and firewood (Wilkinson and Elevitch 2000; Keller 2001; Thomson et al. 2006; Zanan and Nadaf 2011). The fruits and their seeds are consumed, fruit pulp is famine food in many Asian countries. The baked fruit is fragrant, yellowish-orange or reddish-tinged light green (Eland 2012). In the Pacfic, mature drupes of *P. tectorius* are used to stain barkcloth, and as necklaces and garlands; dried seeds are used as paint brushes for painting tapa, for fuel, compost, and as fishing line floats (Wilkinson and Elevitch 2000; Thomson et al. 2006; Eland 2012). In Kiribati, the fruit is also used as bait for catching lobster (Thomson et al. 2006).

The fruits of *P. tectorius* are used as a staple food in parts of Micronesia, the Marshall Islands, the Federated States of Micronesia, and Kiribati, providing up to 50% of energy intake (Miller et al. 1956; Englberger 2003). On average, 100 g *Pandanus* paste provides 321 kcal, 2.2 g protein, 134 mg calcium, 108 mg phosphorus, 5.7 mg iron, 0.04 mg thiamin, 2 mg vitamin C (Murai et al. 1958; Miller et al. 1956; Dignan et al. 1994) and 390–724 μg/100 g beta-carotene (Englberger et al. 2006a, b). In Micronesia, chewing *Pandanus* keys (phalanges) is usually done outside of meal times and is a pleasurable, highly social activity. Adults may typically consume 20–50 keys daily during the main fruiting seasons (Englberger 2003). A 100 g portion of edible pericarp is mainly comprised of water (80 g) and carbohydrates (17 g). There are also significant levels of beta-carotene (19–19,000 μg) and vitamin C (5 mg), and small amounts of protein (1.3 mg), fat (0.7 mg), and fiber (3.5 g) (Englberger 2003; Englberger et al. 2006a, b). *Pandanus* fruit is also a useful source of vitamin C, thiamine, riboflavin, and niacin (vitamin B_3) (Murai et al. 1958; Miller et al. 1956).

Leaves, roots, and flowers are used as folk medicine (Napralert 1993; Wilkinson and Elevitch 2000; Keller 2001; Thomson et al. 2006; Zanan and Nadaf 2011). Tan et al. (2008) observed antitubercular triterpenes and phytosterols activity; diuretic activity has also been shown (Keller 2001). *Pandanus* has high phenolic content, lignin-type compounds, and benzofuran derivatives; methanolic extract of leaves and roots showed strong antioxidant activity (Jong and Chau 1998; Sasikumar et al. 2011) and also showed anti-inflammatory property (Londkar et al. 2010) and antimicrobial activity (Kumar et al. 2010); acetone and ethanol extract showed antifungal activity (Khond et al. 2009). Medicine has employed the bitter-tasting leaves as a purgative and in curing leprosy, syphilis, smallpox, blood disorder, rheumatism, headache, some venereal disease, various skin disorders, scabies, leucoderma, and diseases of the heart and brain, paste of leaves is used on the head to prevent lice (Anonymous 1966; Borua 1995; Khond et al. 2009; Eland 2012). In India, juice from the roots is used to treat a range of problems, from diabetes and fever to sterility and abortion (abortifacent), and to heal ulcers and wounds (Eland 2012). It is used to treat skin diseases, ulcers, dyspepsia, diabetes, fever, and leprosy (Warrier et al. 1995). Staminate flowers are used for earache and blood diseases; spadix juice is used to cure rheumatic arthritis in animals (Anonymous 1966); aromatic oil from bract is a stimulant and antispasmodic and is used to treat rheumatism (Anonymous 1966). Staminate flower and flower oil is used as a stimulant, and it also used to treat headache and rheumatism (Zanan and Nadaf 2011; Eland 2012). The chewed flowers have been given to children by their mothers to swallow as a laxative; the juice of pounded roots has provided an ingredient in potions prepared to ease chest pains and in tonics for women weakened after giving birth (Eland 2012).

Pandanus amaryllifolius

P. amaryllifolius, an herb, is cultivated in home gardens in coastal regions of India; it has a long tradition of cultivation (Routray and Rayaguru 2010). It follows only a vegetative mode of reproduction through lateral buds (Wakte et al. 2010). It is widely used in Southeast Asia for flavoring various food products such as bakery products, sweets, and even home cooking because of its distinct and pleasant aroma (Jiang 1999; Laohakunjit and Noomhorm 2004; Bhattacharjee et al. 2005; Wakte et al. 2007, 2009). The plant bears the same aroma principle 2-acetyl-1-pyrroline (2AP) that has been identified as a principal aroma compound in Basmati and other scented rice (Buttery et al. 1983). In the plant kingdom, *P. amaryllifolius* has the highest-recorded 2AP contents of 14.10 ppm (Thimmaraju et al. 2005). Wakte et al. (2010) identified 31 different volatile compounds from *P. amaryllifolius* leaf, in which 21 are reported for the first time. Recently, an elite population from Palghar locality from Maharashtra state has been identified with the highest 2AP content (12.25 ppm), nonanal (2.97 ppm), and 2-hexenal (25.70 ppm) (Wakte et al. 2012).

There is growing interest in the food industry in natural colorants and flavors eliciting the characteristics of authentic food (Uhl 1995). The Geneva-based International Standards Organization (ISO) has included *P. amaryllifolius* in the document 676 that lists 109 herb and spice plant species useful as ingredients in food (Peter 2001; Wakte et al. 2009). In India, *P. amaryllifolius* leaves are traditionally used for cooking common nonaromatic rice to impart a resemblance of the Basmati aroma. Wakte et al. (2009) has described the plantation techniques. On average, a single mature plant produces 6–10 suckers per year. Leaves are harvested after 6 months of plantation and harvested leaves are tied in bundles for sale or stems along with a bunch of leaves can be directly taken for sale. A one-year-old plant produces about 60 leaves and reaches up to 1 m height in the field (Wakte 2010). It is also used to flavor meat and vegetable products (Ooi et al. 2004).

Powder from the dried leaves has been widely used in ice cream, yogurt, soup, cake, tea, a type of Malaysian traditional pandan-flavored rice, and even Malaysian traditional coconut jam (Che Man et al. 1999; Bhattacharjee et al. 2005). Juices extracted from the leaves are used as an essence in the cake industry and "Nasi lemak" in Malaysia and "Nasi kuning" in Indonesia; a popular breakfast menu is traditionally prepared by cooking the rice with coconut milk ("santan") and *P. amaryllifolius* leaf (Asmain 2010). In India, leaves are traditionally sold in local markets to flavor rice, curries, milk, cakes, puddings, and ice cream (Laohakunjit and Noomhorm 2004; Bhattacharjee et al. 2005; Wakte et al. 2007, 2009). The leaves are added to iced drinks prepared from coconut milk and sweet puddings and custard prepared from sticky, glutinous rice with palm sugar (Wyk 2005). Chopped leaves are mixed with petals of selected scented flowers to make a "potpourri" that is arranged during traditional ceremonies in Malaysia (Samy et al. 2005; Wakte et al. 2009). Wyk (2005) and Wakte et al. (2009) stated that *P. amaryllifolius* essence can be a possible substitute to vanilla essence. Some large-scale cultivation of *P. amaryllifolius* is found in South India to meet the needs of market where it is sold as an additive or spice for flavoring food items; seedlings are usually propagated by detaching rooted suckers or stem cuttings. On an average, 6–10 suckers are produced by an individual plant each year. After separating these plantlets with their aerial roots from the main plant, they can be planted in moist soil by removing most of the leaves but leaving a few at the top. It can be successfully cultivated as an under-storey crop in fruit orchards (Wakte 2010).

Because of high chlorophyll content in *P. amaryllifolius* leaf, it is popularly used as a green colorant in food. The green leaf powder was found to be safe to use as food additive (Nguyen 2006; Porrarud and Pranee 2010). Along with the aromatic properties, the leaves have compounds with antiviral (Ooi et al. 2004) and antioxidant (Nor et al. 2008) properties and alpha-tocopherol content in *P. odorus* (*P. amaryllifolius*) (Ching and Mohamed 2001). This species is also used in some traditional medicine such as remedy for toothache and rheumatism (Takayama et al. 2001a), as a diuretic (Takayama et al. 2002), and as an anti-inflammatory and for decreasing glucose concentration or hypoglycemic effect (Peungvicha et al. 1998a, b), and for curing rheumatic and neuropathy. The root extract is used to cure thyroid problems. In Taiwan, this plant is used to treat fever (Jong and Chau 1998). Tender shoots are directly eaten in case of severe jaundice. The oil obtained from the leaf is described as a stimulant and antispasmodic and is effective against headaches, rheumatism, and epilepsy and as a cure for sore throats (Sankaranarayanan et al. 2010). Busque (2002) reported that *P. amaryllifolius* was used for strengthening heart and as a diuretic. In Thailand, this is a traditional medicine for treating diabetes (Ravindran and Balachandran 2005). Leaves are also used in perfume industry and also medicinally important as diuretic, cardiotonic, antidiabetic and for skin diseases (Keller 2001). The mixture of leaves of *P. amaryllifolius* and *Lantana camara* is a more effective cure for cough (Balagcod and Balangod 2011). In addition, *P. amaryllifolius* is used as a natural and environmentally friendly pest management tool, for example, taxi drivers in Singapore and Malaysia keep bunches of *P. amaryllifolius* in their taxis to ward off cockroaches (Samy et al. 2005). Leaves of *P. amaryllifolius* are used as perfume (Nguyen 2006). Traditionally, leaves are used as a medicinal bath for women after childbirth in Malaysia and also for hair washing. It is also used for preparing lotion along with ash and vinegar to treat measles, as a purgative, in the treatment of leprosy, for sore throat and as a diuretic in Philippines (Samy et al. 2005). Traditionally, a mixture of henna (*Lawsonia inermis*), limau purut (kaffir lime, or *Citrus hystrix*), coconut milk, milk, and *P. amaryllifolius* leaf is used to clean hair and to provide fragrance (Turner 2007).

Hot water extracts of the roots of *P. amaryllifolius* showed hypoglycemic activity and 4-hydroxybenzoic acid has been isolated as the active principle (Peungvicha et al. 1996, 1998a, b). The plant is also an important source of several alkaloids, such as(+)-Pandamarine, a piperidine-type alkaloid with lactam moieties (Byrne et al. 1992); Pandamarilactone-1, Pandamarilactone-32, and Pandamarilactone-31 are also piperidine type but with lactone instead of lactam moieties (Nonato et al. 1993); Pandamarilactam-3x and Pandamarilactam-3y are pyrrolidinone-type alkaloids with lactone moiety (Sjaifullah and Garson 1996); Pandamarilactonine-A, -B, -C, -D and Norpandamarilactonine-A, -B are pyrrolidine-type alkaloids with lactone moiety, and 6Z-Pandanamine, a symmetrical secondary amine with lactone moieties

(Takayama et al. 2000, 2001a, b, 2002); 6E-Pandanamine and artifacts (Salim et al. 2004); pandanin (Ooi et al. 2004).

Pandanus leram

In the Andaman and Nicobar Islands, *P. leram* is located in the interior of littoral forest of Rutland Island, having large-size fruits and high pulp content, it is used as a staple food (about 1.6%), and dried fruits are used as brush (Sharif and Rao 2007; Abraham et al. 2008; UNESCO 2010). The fruit is a staple food of "*Shompen*" tribes; they boil the trimmed cones for 5–8 h (depending on the state of ripeness of the fruit), remove fibrous material, which yields a yellow paste after processing that is consumed. They make the paste daily, preserving it for a short time by packing it in large green leaves in the form of round balls. The "*Shompen*" are nomadic in nature and they follow a set pattern of moving from one *Pandanus* tree grove to another in a more or less predetermined fashion based on the yield of the grove (Patnaik and Prasad 2009). Katiyar et al. (1989) reported that the fruit pulp and seeds contain 75.8 and 57.1% moisture, 0.6 and 0.9% total mineral content, 0.4 and 7.1% protein, 8.1 and 3.3% fiber, 0.5 and 23.7% total lipid, and 14.6 and 7.9% non-fiber carbohydrates, respectively. The seeds are nutritionally rich in comparison to the fruit pulp, but are available only in small fraction (3%) of the total fruit. They also observed major fatty acids in the seed oil (palmitic – 56.4%, oleic – 26.5%, and linoleic acids – 16.4%). The Negrito tribes of the Andaman Islands use leaves for treatment of body pain (Sharif 2007). In the Jarawa territory, young fronds of *Angiopteris lygodiifolia* and midrib of leaves of *Pandanus andamanensis* are tied around the chest to control cough, cold, and body pain (http://www.and.nic.in).

Pandanus dubius

In India, *P. dubius* is used as ornamental plant (Keller 2001). In Guam, leaves are useful for waterproof matting, preparation of bags, coarse wicker-work, and making umbrellas; seeds are edible with a similar taste to that of coconut (Wilkinson and Elevitch 2000; Keller 2001; Yoshioka 2006). Young seeds have medicinal uses in case of food poisoning (Keller 2001). Tan et al. (2010) isolated for the first time two new alkaloids, dubiusamine A and dubiusamine B, from the leaves.

Pandanus unipapillatus

In India, *P. unipapillatus* is used as a hedge plant and soil binder. In Kerala, live plants are used to prevent entry of wild animals in fields, leaves are used for weaving mats and baskets, and fruit is used as a food stuff (Prasad and Raveendran 2010; Zanan and Nadaf 2011). Kumar (2011) reported that it is also used for medicinal purposes.

Pandanus emarginatus

P. emarginatus is found in the wild forests of Arunachal Pradesh in India. Its fruits are used as a food stuff, the leaves are used by local tribes for weaving mats and production of fiber; different plant parts are used in folk medicine.

Pandanus nepalensis

In India, *P. nepalensis* is grown in the Sikkim and West Bengal states (Joshi and Joshi 1991). The fresh leaves act as a cockroach repellant. The leaves are used for making handwoven mats, carrying bags, fishing bags, thatching roofs, and as dyes. It is chewed as a breath sweetener. The fruits are edible, and are eaten by monkeys and rats. It has som medicinally important properties: the Lepcha tribe from the Dzongu Valley in North Sikkim uses leaves as an antidote to snake poison; young or tender leaves are placed on the skin at the site of snake bite to reduce the pain (Pradhan and Badola 2008; Sundriyal et al. 2004).

Pandanus furcatus

In India, *P. furcatus* is mainly distributed in the Malabar region. It is used as a hedge plant and soil binder; stems are used for preparing floats and fishing nets, leaves are used for weaving mats and baskets, and fruit is used as a food stuff. It acts as a remedy for snake bite (Keller 2001; Sujana and Sivaperuman 2007; Zanan and Nadaf 2011).

Pandanus kaida

P. kaida is distributed in the Malabar region of India. The plant is used as a hedge plant and soil binder; the plant canopy is used for water filter and purification. In Kerala, live plants are used to prevent entry of wild animals into fields; leaves are used for weaving mats and baskets (Melvani 2005; Prasad and Raveendran 2010; Zanan and Nadaf 2011). Floral bracts are kept along with clothes to impart the fragrance (Zanan and Nadaf 2011).

Pandanus unguifer

P. unguifer is the only *Pandanus* species that produces flowers in potted condition and hence, has ornamental value (Srivastava and Choudhary 2007).

Pandanus martinianus

P. martinianus is found in low land area and hilly regions of Assam and Arunachal Pradesh. It is used as hedge plant around the paddy field, fragrant flowers are used for fragrance and fruits are consumed by local people (Zanan and Nadaf 2012b).

Pandanus diversus

P. diversus is found in the wild in forests in the Assam state of India. Its fruits are used as food a stuff and the leaves are used by local tribes for production of fiber.

Pandanus palakkadensis and *Pandanus mangalorensis*

P. palakkadensis and *P. mangalorensis* are used as a hedge plant along the boundaries of paddy fields to protect the crop (Nadaf et al. 2011; Zanan and Nadaf 2012a).

Benstonea thwaitesii

B. thwaitesii is an evergreen species found in wet evergreen forest (Ramesh et al. 2009). Leaves are used for weaving mats, brushes, and diverse domestic articles and the fruit pulp is edible (Umapathy 1998; Krishnamani and Kumar 2000; Keller 2001; Zanan and Nadaf 2011). Floral bracts are offerings in religious ceremonies and are kept along with clothes to impart their fragrance (Zanan and Nadaf 2011).

Benstonea foetida

In India, *B. foetida* is used as a hedge plant; leaves are used for matting, paper making, and thatching; fiber from leaf is used for nets and brushes; leaf juice is used on the face; and the fruit is edible (Prain 1963; Bennet 1979; Dey 1995; Zanan and Nadaf 2011). The whole plant is used in traditional medicine for leprosy, small pox, syphilis, scabies, and heart and brain diseases; leaves are used for spadix and diabetes (Uddin et al. 2011). Uddin et al. (2006) observed antinociceptive activity in methanolic extract of leaf.

References

Abraham Z, Senthilkumar R, John KJ, Sharma TVRS, Nair NV, Unnikrishnan M, Kumaran PM, George JK, Uma S, Latha M, Malik SS, Mishra SK, Bhandari DC, Pareek SK (2008) Collection of plant genetic resources from Andaman and Nicobar islands. Genet Resour Crop Evol 55:1279–1289. doi:10.1007/s10722-008-9327-9

Anonymous (1966) Wealth of India, raw materials, vol 3. Publications and Information Directorate, CSIR, New Delhi

Anonymous (1996) A Survey report on Kewda in Ganjam district. Regional Research Laboratory (CSIR), Bhubaneswar

Asmain AB (2010) Extracts of the subterranean root from *Pandanus amaryllifolius*. Dissertation for the Bachelor of Pharmacy, Universiti Teknologi Mara

Balagcod TD, Balangod AKD (2011) Ethnomedical knowledge of plants and healthcare practices among the Kalanguya tribe in Tinoc, Ifugao, Luzon, Philippines. Indian J Trad Knowl 10(2):227–238

Bennet SSR (1979) Flora of Howrah district. International Book Distributor, Dehra Dun, pp 23–24

Bhattacharjee P, Kshirsagar A, Singhal RS (2005) Supercritical carbon dioxide extraction of 2-acetyl-1-pyrroline from *Pandanus amaryllifolius* Roxb. Food Chem 91:255–259. doi:10.1016/j.foodchem.2004.01.062

Borua B (1995) *Gachpala Tarulata* (in Bangla). In: Bhuyan AI (ed) Plants and herbs. Bangla Academy, Dhaka, Bangladesh, p 200

Busque F, March P, Figueredo M, Font J, Sanfeliu E (2002) Total synthesis of four *Pandanus* alkaloids: pandamarilactonine-A and -B and their chemical precursors norpandamarilactonine-A and -B. Tetrahedron Lett 43:5583–5585. doi:doi.org/10.1016/S0040-4039(02)01129-2

Buttery RG, Juliano BO, Ling LC (1983) Identification of rice aroma compound 2-acetyl-1-pyrroline in pandan leaves. Chem Ind 23:478

Byrne LT, Guevara BQ, Patalinghug WC, Recio BB, Ualat CR, White AH (1992) The X-ray crystal structure of (+) Pandamarine, the major alkaloid of *Pandanus amaryllifolius*. Aust J Chem 45:1903–1908

Callmander MW (2001) *Pandanus* subg. *Martellidendron* (Pandanaceae) part II: revision of sect. *Martellidendron* Pic. Serm. in Madagascar. Bot J Linn Soc 137:353–374

Callmander MW, Chassot P, Küpfer P, Lowry PP II (2003) Recognition of *Martellidendron*, a new genus of Pandanaceae, and its biogeographic implications. Taxon 52(4):747–762

Che Man YB, Irwandi J, Abdullah WJW (1999) Effect of different types of maltodextrin and drying methods on physicochemical and sensory properties of encapsulated durian flavour. J Sci Food Agric 79:1075–1080
Ching LS, Mohamed S (2001) Alpha-Tocopherol content in 62 edible tropical plants. J Agric Food Chem 49:3101–3105
Clayton N (2007) Animal cognition: crows spontaneously solve a metatool task. Curr Biol 17(20):R894–R895
Dey TK (1995) *Bangladesher Proyojonio Gachgacgra* (in Bangla) (useful plants of Bangladesh). Comilla. Shetu Tuni Book House, Bangladesh, p 387
Dignan CA, Burlingame BA, Arthur JM, Quigley RJ, Milligan GC (1994) The Pacific Islands food composition tables. South Pacific Commission, Noumea
Dutta PK, Saxana HO, Brahmam M (1987) Kewda perfume industry in India. Econ Bot 41:403–410
Eland S (2012) The plant biographies. http://www.plantlives.com
Englberger L (2003) Are Pacific islanders still enjoying the taste of *Pandanus*? Pac Islands Nutr 58:10–11
Englberger L, Aalbersberg W, Dolodolotawake U, Schierle J, Humphries J, Iuta T, Marks GC, Fitzgerald MH, Rimon B, Kaiririete M (2006a) Carotenoid content of *Pandanus* fruit cultivars and other foods of the Republic of Kiribati. Public Health Nutr 9(5):631–643
Englberger L, Aalbersberg W, Schierle J, Marks GC, Fitzgerald MH, Muller F, Jekkein A, Alfred J, Van der Velde N (2006b) Carotenoid content of different edible *Pandanus* fruit cultivars of the Republic of the Marshall Islands. J Food Comp Anal 19(6–7):484–494
http://www.and.nic.in
Hunt GR, Gray RD (2004) Direct observations of Pandanus-tool manufacture and use by a New Caledonian crow (*Corvus moneduloides*). Anim Cogn 7:114–120
Jiang J (1999) Volatile composition of pandan leaves (*Pandanus amaryllifolius*). In: Shadidi F, Ho CT (eds) Flavour chemistry of ethnic foods. Kluwer, New York, pp 105–109
Jong TT, Chau SW (1998) Antioxidative activities of constituents isolated from *Pandanus odoratissimus*. Phytochemistry 49:2145–2148. doi:doi.org/10.1016/S0031-9422(98)00390-2
Joshi AR, Joshi DP (1991) Endemic plants of Nepal Himalaya: conservation status and future direction. Mount Environ Dev 1(2):1–35
Katiyar SK, Kumar N, Bhatia AK (1989) A chemical study of *Pandanus leram* fruit grown in the Andaman and Nicobar islands. Trop Sci 29(2):137–140
Keller J (2001) Pandanaceae. In: Hanelt P, Institute of Plant Genetics and Crop Plant Research (eds) Mansfeld's encyclopedia of agricultural and horticultural crops, vol 5. Springer, Berlin, pp 2816–2824
Khond M, Bhosale JD, Tasleem A, Mandal TK, Padhi MM, Dabur R (2009) Screening of some selected medicinal plants extracts for in-vitro antimicrobial activity. Middle East J Sci Res 4(4):271–278
Krishnamani R, Kumar A (2000) Phyto-ecology of the lion-tailed macaque (*Macaca silenus*) habitats in Karnataka, India: floristic structure and density of food-trees. Primate Rep 58:27–56
Kumar BM (2011) Species richness and aboveground carbon stocks in the homegardens of central Kerala, India. Agric Ecosyst Environ. doi:10.1016/j.agee.2011.01.006
Kumar D, Kumar S, Kumar S, Singh J, Sharma C, Aneja KR (2010) Antimicrobial and preliminary phytochemical screening of crude leaf extract of *Pandanus odoratissimus* L. Pharmacol Online 2:600–610
Laohakunjit N, Noomhorm A (2004) Supercritical carbon dioxide extraction of 2-acetyl-1-pyrroline and volatile components from pandan leaves. Flavour Frag J 19:251–259. doi:10.1016/j.foodchem.2004.01.062
Londkar R, Kamble A, Reddy C (2010) Anti-inflammatory activity of *Pandanus odoratissimus* extract. Int J Pharmacol 6(3):311–314
McKenzie E, Whitton S, Hyde K (2001) The Pandanaceae – does it have a diverse and unique fungal biota? Abstract in Australian Mycological Society Conference
Melvani K (2005) The role of trees in the bioremediation of drinking water in Nawakkaduwa, Sri Lanka. In: Ploeg J, Masipiqueña AB (eds) The future of the Sierra Madre: responding to social and ecological changes, 3rd edn. Wiley, New York, pp 35–54
Miller CD, Murai M, Pen F (1956) The use of *Pandanus* fruit as food in Micronesia. Pac Sci 10:3–16
Mohsin HF, Wahab AI, Naz H, Shah ASA, Weber JFF (2008) Pandanaceae research in support for Industrial Plantation. In: 2nd international plantation industry conference and exhibition, Malaysia. November 18–21, 2008
Murai M, Pen F, Miller CD (1958) Some tropical South Pacific Island foods. Description, history, use, composition, and nutritive value. University of Hawaii Press, Honolulu
Nadaf AB, Zanan RL, Wakte KV (2011) A new endemic species of Pandanaceae from India: *Pandanus palakkadensis*. Kew Bull 66:1–5
Napralert x (1993) Biological activities for extracts of *Pandanus* and ethnomedical information on *Pandanus*. University of Illinois at Chicago, Chicago
Nguyen MLT (2006) Insertions and deletions: evolution in the assemblage of Vietnamese food plants. Ethnobot Res Appl 4:175–201
Nonato MG, Garson MJ, Truscott RJW, Carver JA (1993) Structural characterization of piperidine alkaloids from *Pandanus amaryllifolius* by inverse-detected 2D NMR techniques. Phytochemistry 34:1159–1163

Nor FM, Mohamed S, Idris NA, Ismail R (2008) Antioxidative properties of *Pandanus amaryllifolius* leaf extracts in accelerated oxidation and deep frying studies. Food Chem 110:319–327. doi:10.1016/j.foodchem.2008.02.004

Novellino D, Ertuğ ZF (2005) Baskets of the world, the social significance of plaited crafts. In: Proceedings of the IVth international congress of ethnobotany (ICEB 2005), Yeditepe University, Istanbul, 21–26 Aug, 2005

Ooi LSM, Sun SSM, Ooi VEC (2004) Purification and characterization of a new antiviral protein from the leaves of *Pandanus amaryllifolius* (Pandanaceae). Int J Biochem Cell Biol 36:1440–1446. doi:doi.org/10.1016/j.biocel.2004.01.015

Panda KK, Mohapatra S, Das LN, Misra MK, Panda BB (2001) Optimal utilization of kewda *Pandanus fascicularis* to ameliorate economy and ecology of coastal India. J Med Aromat Plant Sci 22–23:679–682

Panda KK, Panigrahi RK, Das AB, Panda BB (2007) Analyses of chromosome number, nuclear DNA content and RAPD profile in three morphotypes of *Pandanus fascicularis* Lam. Plant Genet Resour Newsl 152:12–22

Panda KK, Das AB, Panda BB (2009) Use and variation of *Pandanus tectorius* Parkinson (*P. fascicularis* Lam.) along the coastline of Orissa, India. Genet Resour Crop Evol 56:629–637. doi:10.1007/s10722-008-9390-2

Patnaik R, Prasad BVR (2009) Ethnic identity of the indigenous tribes in Great Nicobar island. Stud Tribes Tribals 7(2):119–123

Peter KV (2001) Handbook of herbs and spices. Woodhead Publishing Ltd, Cambridge, pp 26–27

Peungvicha P, Thirwapan SS, Watanabe H (1996) Hypoglycaemic water extract of root of *Pandanus odorus* RIDL. Biol Pharm Bull 19(3):364–366

Peungvicha P, Temsiririrkkul R, Prasain JK, Tezuka Y, Kadota S, Thirawarapan SS, Watanabe H (1998a) 4-Hydroxybenzoic acid: a hypoglycemic constituent of aqueous extract of *Pandanus odorus* root. J Ethnopharmacol 62(1):79–84. doi:doi.org/10.1016/S0378-8741(98)00061-0

Peungvicha P, Thirawarapan SS, Watanabe H (1998b) 4-Hydroxybenzoic acid: a hypoglycemic constituent of aqueous extract of *Pandanus odorus* root II. Jpn J Pharmacol 78:395–398. doi:10.1254/jjp. 78.395

Porrarud S, Pranee A (2010) Microencapsulation of Zn-chlorophyll pigment from Pandan leaf by spray drying and its characteristic. Int Food Res J 17:1031–1042

Pradhan BK, Badola HK (2008) Ethnomedicinal plant use by Lepcha tribe of Dzongu valley, bordering Khangchendzonga biosphere reserve, in north Sikkim, India. J Ethnobiol Ethnomed 4(22):1–18. doi:10.1186/1746-4269-4-22

Prain D (1963) Bengal plants. Botanical Survey of India, Calcutta, pp 828–829

Prasad SK, Raveendran K (2010) Traditional plant fencing and its conservatory nature in Kasaragod district, Kerala, India. Ethnobot Leaflets 14:681–686

Raina VK, Kumar A, Srivastava SK, Syamsunder KV, Kahol AP (2004) Essential oil composition of 'kewda' (*Pandanus odoratissimus*) from India. Flavour Frag 19:434–436. doi:10.1002/ffj.1331

Ramesh BR, Swaminath MH, Patil S, Aravajy S, Elouard C (2009) Pondy papers in ecology no. 7: assessment and conservation of forest biodiversity in the Western Ghats of Karnataka, India. 2. Assessment of tree biodiversity, logging impact and general discussion. French Institute Pondicherry

Ravindran PN, Balachandran I (2005) Underutilized medicinal species-III. Spice India 18(2):16–24

Routray W, Rayaguru K (2010) Chemical constituents and post-harvest prospects of *Pandanus amaryllifolius* leaves: a review. Food Rev Int 26(3):230–245. doi:10.1080/87559129.2010.484114

Sahu D, Misra MK (2007) Ecology and traditional technology of screw pine perfume industry in Coastal Orissa. Indian J Trad Knowl 6:253–261

Salim AA, Garson MJ, Craik DJ (2004) New alkaloids from *Pandanus amaryllifolius*. J Nat Prod 67:54–57

Samarawickrema WA, Sone F, Self LS, Cimmings RF, Paulson GS (2008) Distribution of breeding and control of the filariasis vector *Aedes samoanus* in leaf axils of *Pandanus* in Samoa. Med Vet Entomol 6(4):367–370

Samy J, Sugumaran M, Kate LWL (2005) Herbs of Malaysia: an introduction to the medicinal, culinary, aromatic and cosmetic use of herbs. Federal Publications, Times Editions-Marshall Cavendish, Malaysia, pp 180–181

Sankaranarayanan S, Bama P, Ramachandran J, Kalaichelvan PT, Deccaraman M, Vijayalakshimi M, Dhamotharan R, Dananjeyan B, Bama S (2010) Ethnobotanical study of medicinal plants used by traditional users in Villupuram district of Tamil Nadu, India. J Med Plant Res 4(12):1089–1101

Sasikumar JM, Jinu U, Shamna R (2011) In vitro antioxidant activity and HPTLC analysis of root of *Pandanus odoratissimus* L. Int J Pharm Pharm Sci 3(1):64–67

Sharif MU (2007) Plants folk medicine of Negrito tribes of Bay Islands. Indian J Trad Knowl 6(3):468–476

Sharif MU, Rao RR (2007) Ethnobotanical studies of Shompens – a critically endangered and degenerating ethnic community in Great Nicobar island. Curr Sci 93(11):1623–1628

Sjaifullah A, Garson MJ (1996) Structural characterization of two novel pyrrolidinones from *Pandanus amaryllifolius* Roxb. ACGC Chem Res Commun 5:24–27

Srivastava RC, Choudhary RK (2007) Rediscovery of *Pandanus unguifer* Hook. f. (Pandanaceae): a RET species from its type-locality after 75 years. Curr Sci 92(10):1325

St. John H (1963) The proposal (93) to conserve *Pandanus* L. f. Taxon 12(5):201–204

Stone BC (1983) A guide to collecting Pandanaceae (*Pandanus, Freycinetia* and *Sararanga*). Ann Mo Bot Gard 70(1):137–145

Sujana KA, Sivaperuman C (2007) Useful macrophytes in the vembanad kole Ramsar Site, Kerala. Indian J Environ Educ 7:27–37

Sundriyal M, Sundriyal RC, Sharma E (2004) Dietary use of wild plant resources in the Sikkim Himalaya, India. Econ Bot 58(4):626–638

Takayama H, Ichikawa T, Kuwajima T, Kitajima M, Seki H, Aimi N, Nonato MG (2000) Structure characterization, biomimetic total synthesis, and optical purity of two new pyrrolidine alkaloids, pandamarilactonine-A and -B, isolated from *Pandanus amaryllifolius* Roxb. J Am Chem Soc 122(36):8635–8639. doi:10.1021/ja0009929

Takayama H, Ichikawa T, Kitajima M, Aimi N, Lopez D, Nonato MG (2001a) A new alkaloid, pandanamine, finding of an anticipated biogenetic intermediate in *Pandanus amaryllifolius* Robx.. Tetrahedron Lett 42:2995–2996. doi:10.1016/S0040-4039(01)00339-2

Takayama H, Ichikawa T, Kitajima M, Nonato MG, Aimi N (2001b) Isolation and characterization of two new alkaloids, Norpandamarilactonine-A and –B, from *Pandanus amaryllifolius* by spectroscopic and synthetic methods. J Nat Prod 64:1224–1225

Takayama H, Ichikawa T, Kitajima M, Nonato MG, Aimi N (2002) Isolation and structure elucidation of two new alkaloids, pandamarilactonine-C and -D, from *Pandanus amaryllifolius* and revision of relative stereochemistry of Pandamarilactonine-A and -B by total synthesis. Chem Pharm Bull(Tokyo) 50(9):1303–1304. doi:10.1248/cpb.50.1303

Takeda J, De Silva S, Muthuraman P, Rahman SM, Kawet L (2008) Spices in Sri Lanka, India and Bangladesh with special reference to the usages and consumptions. Agri Bull 93:1–25

Tan MA, Takayama H, Aimi N, Kitajima M, Franzblau SG, Nonato MG (2008) Antitubercular triterpenes and phytosterols from *Pandanus tectorius* Soland. var. *laevis*. J Nat Med 62(2):232–235. doi:10.1007/s11418-007-0218-8

Tan MA, Kitajima M, Kogure N, Nonato MG, Takayama H (2010) Isolation and total syntheses of two new alkaloids, dubiusamines-A, and -B, from *Pandanus dubius*. Tetrahedron 66:3353–3359

Tanaka N, Sasaki Y, Mowjood MIM, Jinadasa KBSN, Homchuen S (2007) Coastal vegetation structures and their functions in tsunami protection: experience of the recent Indian Ocean tsunami. Landscape Ecol Eng 3(1):33–45. doi:10.1007/s11355-006-0013-9

Tanaka N, Nandasena NAK, Jinadasa KBSN, Sasaki Y, Tanimoto K, Mowjood MIM (2009) Developing effective vegetation bioshield for tsunami protection. Civ Eng Environ Syst 26(2):163–180. doi:10.1080/10286600802435850

Tanaka N, Jinadasa KBSN, Mowjood MIM, Fasly MSM (2011) Coastal vegetation planting projects for tsunami disaster mitigation: effectiveness evaluation of new establishments. Landscape Ecol Eng 7:127–135. doi:10.1007/s11355-010-0122-3

Thimmaraju R, Bhagyalakshmi N, Narayan MS, Venkatachalam L, Ravishankar GA (2005) In vitro culture of *Pandanus amaryllifolius* and enhancement of 2-acetyl-1-pyrroline, the major flavouring compound of aromatic rice, by precursor feeding of L-proline. J Sci Food Agric 85(15):2527–2534. doi:10.1002/jsfa.2286

Thomson LAJ, Englberger L, Guarino L, Thaman, Elevitch CR (2006) *Pandanus tectorius* (Pandanus), ver. 1.1. In: Elevitch CR (ed) Species profiles for Pacific Island agroforestry. Permanent Agriculture Resources (PAR), Holualoa. http://www.traditionaltree.org

Thuy NB, Tanimoto K, Tanaka N, Harada K, Iimura K (2009) Effect of open gap in coastal forest on tsunami run-up-investigations by experiment and numerical simulation. Ocean Eng 36(15–16):1258–1269

Turner I (2007) Plants for your hair-henna and company. Gardenwise, The Newsletter of the Singapore Botanic Gardens. 28:36

Uddin SJ, Shilpi JA, Rahman MT, Ferdous M, Rouf R, Sarker SD (2006) Assessment of neuropharmacological activities of *Pandanus foetidus* (Pandanaceae) in mice. Pharmazie 61(4):362–364

Uddin SJ, Grice ID, Tiralongo E (2011) Cytotoxic effects of Bangladeshi medicinal plant extracts. Evidence-based complementary and alternative medicine, ID 578092. doi:10.1093/ecam/nep111

Uhl S (1995) Emerging flavors of global village-growing demand for ethnic food. Prepared Foods. Food & Beverage Industry

Umapathy G (1998) Impacts of habitat fragmentation on the arboreal mammals in the wet evergreen forests of the Anaimalai Hills in the Western Ghats, south India. Ph.D. thesis, Bharathiar University, Coimbatore

UNESCO (2010) The Jarawa Tribal reserve dossier: cultural & biological diversities in the Andaman islands. In: Sekhsaria P, Pandya V (ed) UNESCO, Paris, 212 p

Wakte KV (2010) Biochemical characterization of aroma volatiles and molecular genotyping in Indian *Pandanus amaryllifolius* Roxb. clones. Dissertation, University of Pune

Wakte KV, Nadaf AB, Krishnan S, Thengane RJ (2007) Studies on lower epidermal papillae, site of storage of "basmati rice" aroma compounds in *Pandanus amaryllifolius* Roxb. Curr Sci 93(2):238–242

Wakte KV, Nadaf AB, Thengane RJ, Jawali N (2009) *Pandanus amaryllifolius* Roxb. cultivated as a spice in coastal regions of India. Genet Resour Crop Evol 56:735–740. doi:10.1007/s10722-009-9431-5

Wakte KV, Thengane RJ, Jawali N, Nadaf AB (2010) Optimization of HS-SPME conditions for quantification of 2-acetyl-1-pyrroline and study of other volatiles in *Pandanus amaryllifolius* Roxb. Food Chem 121:595–600. doi:10.1016/j.foodchem.2009.12.056

Wakte KV, Zanan RL, Thengane RJ, Jawali N, Nadaf AB (2012) Identification of elite population of *Pandanus amaryllifolius* Roxb. for higher 2-acetyl-1-pyrroline and other volatile contents by HS-SPME/GC-FID from Peninsular India. Food Anal Method 5:1276–1288. doi:10.1007/s12161-012-9373-y

Warrier RK, Nambiar VPK, Ramankutty C (1995) Indian medicinal plants: a compendium of 500 species. Orient Longman Ltd, Madras

Wilkinson KM, Elevitch CR (2000) Nontimber forest products for Pacific Islands: an introductory guide for producers. In: Elevitch CR, Wilkinson KM (eds) Agroforestry guides for Pacific Islands. Permanent Agriculture Resources, Holualoa

Wyk BEV (2005) Food plants of the world: identification, culinary uses and nutritional value. Times Editions–Marshall Cavendish, Singapore, p 275

Yoshioka J (2006) Limestone forests – botanical gems (Guam). Pac Island Netw News 4:6

Zanan RL, Nadaf AB (2011) Collection, characterization and olfactory evaluation of Pandanus species in southern India. Plant Genet Resour 1–3. doi:10.1017/S1479262111000748

Zanan RL, Nadaf AB (2012a) *Pandanus mangalorensis*: a new species of Pandanaceae from southern India. Kew Bull 67(3):555–559

Zanan RL, Nadaf AB (2012b) *Pandanus martinianus* (Pandanaceae), a new species endemic from North Eastern region of India. Phytotaxa 73:1–7

Chapter 8
Conservation of Indian Pandanaceae: Current Status and Need

Conservation is meant to protect and to make sustainable the use of available genetic resources. The genetic resources play an important role in environmental protection, sustainable management, and conservation of biodiversity. Conservation of genetic resources not only prevents extinction of a species but also reduction of gene pool and loss of genes and genotypes (Nghia 2001). Deforestation and shifting cultivation are the main causes of forest fragmentation, which can lead to a decline of natural populations and a loss of genetic diversity (Nghia 2001). During the past five decades, extensive human activities such as livestock grazing, wood fuel cutting, urbanization, and temporary arid land cultivation have put a great pressure on vegetation and led to vegetation deterioration (Heady 1963; Batanouny 1991). Conservation of plant diversity assumes greater importance when the world is facing an unprecedented loss of biological diversity. A number of approaches for setting priorities have been developed and used by various agencies worldwide; however, the IUCN red listing is probably the most widely used and accepted approach for prioritizing species at the global level (http://www.hcvnetwork.org). As per an estimate, about 60,000 out of 287,655 species of plants known in the world are facing the threat of extinction. As per the revised 1994 IUCN Red List Categories; among the 11,824 species that were evaluated for their threat status, 8,321 species are now on the IUCN Red List 2004 (Baillie et al. 2004).

Indian as a Mega Diversity Centre

The tropical and subtropical countries that are characterized by high species diversity and a greater number of endemic species are identified as mega diverse countries. India is one of the 12 M biodiversity centers of the world and rich in biological diversity, associated with the traditional and contemporary knowledge system related thereto (Biological diversity Act 2002). India shares 2.4% of the land area and accounts for 7–8% of the recorded species of the world, harboring 47,000 species of plants and 89,000 species of animals (Khoshoo 1996; MoEF 2009). About 33% of the flowering plants recorded are endemic to India. There are 16 major forest types in India and it is also one of the eight primary centers of the origin of cultivated plants and is rich in agricultural biodiversity. The most pronounced endemic areas within the country are the Northeast, Western Ghats, Northwest Himalayas, and Andaman and Nicobar Islands. About 60% of the total area has bee explored so far and the remaining 40% has the potential to add to the total list of endemic species (Singh et al. 2008). The Kerala state can be taken as one of the examples from which 250 new species of flowering plants have been reported in the past 25 years (Manilal 2001). According to the 2003 Red List, in India, 45 species are critically endangered, 113 endangered, and 88 vulnerable (Singh et al. 2008).

A. Nadaf and R. Zanan, *Indian Pandanaceae - an overview*,
DOI 10.1007/978-81-322-0753-5_8, © Springer India 2012

India harbors 2 of the 34 hot spots of the world that are characterized by a high degree of endemism and are therefore areas of global conservation concern. These are the Western Ghats and Eastern Himalayas, extending to the neighboring countries (http://www.conservation.org/where/priority_areas/hotspots/hotspots_revisited/Pages/hotspots_revisited.aspx).

Eastern Himalaya

Phytogeographically, the Eastern Himalayas includes a distinct floral region comprising of Nepal, Bhutan, the states of East and Northeast India ,and a contiguous sector of the Yunnan province in Southwestern China. In the whole of Eastern Himalaya, there are an estimated 9,000 plant species, out of which 3,500 (39%) are endemic. In the Indian sector, about 5,800 plant species are recorded that include 2,000 (36%) endemic species. At least 55 flowering plants endemic to this area are recognized as rare. The area has long been recognized as a rich center of primitive flowering plants and is popularly known as the "Cradle of Speciation." Species of several families of monocotyledons, Orchidaceae, Zingiberaceae, and Arecaceae are found in the area (MoEF 2009).

Western Ghats

The Western Ghats region, which is spread into six states of India, is considered to be one of the most important bio-geographic zones of India, as it is one of the richest centers of endemism. Due to varied topography and microclimatic regimes, some areas within the region are considered to be active zones of speciation. As per the MoEF status report (2009), the region has 490 arborescent taxa, of which as many as 308 are endemic. About 1,500 endemic species of dicotyledonous plants are reported from the Western Ghats. Two hundred forty-five species of orchids belonging to 75 genera are found here, of which 112 species in 10 genera are endemic to the region. Many of the endemic and other species are listed as threatened. Nearly 235 species of endemic flowering plants are considered endangered.

Conservation Status of Indian Pandanaceae

Indian Pandanaceae is distributed in the Northeastern Himalayan region and Western Ghats. Traditionally, most of the *Pandanus* species are extensively used for folk medicine, fiber production, oil production, and as food. Because of extensive use by tribal peoples, urbanization, construction, and cultivation practices, many populations of *Pandanus* species are reduced. Some *Pandanus* species are already threatened, the population sizes of some species are severely reduced, and species become scarce. This has developed into an alarming situation, and has registered an urgent need to conserve family Pandanaceae. The value for the conservation and conservation status of Indian *Pandanus* and *Benstonea* species has been not attempted so far. In order to address this challenge and to arrive at an efficient strategy that helps to prioritize the species for an appropriate conservation measure, we conducted a threat assessment exercise following IUCN Red List Categories and Criteria in different states of the country during 2007–2012.

Assessment of Indian Pandanaceae Using the IUCN Red List Categories and Criteria (2001)

The IUCN Red List is a catalogue of taxa that are facing the risk of extinction. The World Conservation Union (IUCN, formerly known as International Union for the Conservation of Nature and Natural Resources) has recognized nine Red List Categories of species: Extinct (EX), Extinct in the Wild (EW), Critically Endangered (CR), Endangered (EN), Vulnerable (VU), Near Threatened (NT), Least Concern (LC), Data Deficient (DD) and Not Evaluated (NE) (IUCN 2001). Callmander et al. (2007), while discussing the identification of priority areas for plant conservation in Madagascar using Red List criteria for rare and threatened Pandanaceae, mentioned that for the countries where detailed inventorization is being carried out, instead of depending only upon the specimen data estimated using a grid cell size, it should be complemented with field observations to determine population size, sensitivity to disturbance, and specific threats to habitat and therefore potential population decline. This will help the conservation planning process. Therefore, in the present study, the IUCN status of the threatened species has been determined following the methodology suggested by Callmander et al. (2007). The details are as follows. Based on a survey, localities for respective species were identified, a grid of 3×3 km was placed, and the numbers of populations and subpopulations were determined. Based on this data, extent of occurrence (EOO, minimum convex polygon containing all points of occurrence) and area of occupancy (AOO, area estimated by superimposing a grid onto occurrence points and calculating the cumulative area of cells occupied by a species) were determined, and the status was determined. Species-wise status is given below.

Pandanus furcatus

Pandanus furcatus predominantly grows in the Malabar region of India. As per the IUCN Red List criteria (IUCN 2001), the species fall under the Least Concern (LC) category.

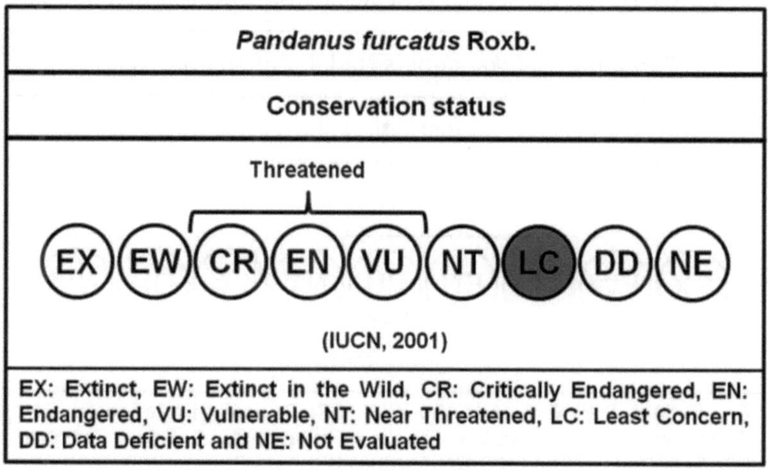

Pandanus unipapillatus

Pandanus unipapillatus is predominantly known in the Malabar region of India. As per the IUCN Red List criteria (IUCN 2001), the species is categorized under the Least Concern (LC) category.

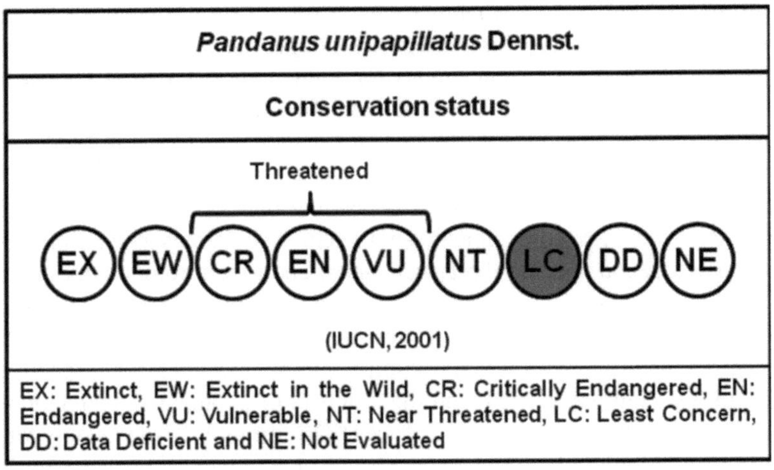

Pandanus palakkadensis

The Palakkad district harbors a high number of endemic species. *P. palakkadensis* is restricted to the type locality and, following the IUCN Red List Categories and Criteria (IUCN 2001), the species is given a preliminary conservation assessment of Critically Endangered (CR 2ab [i, ii, iii, iv]). The locality is affected by shifting cultivation practices and is also used for the rearing of ducks for which plants are being removed. Hence, the species is in urgent need of conservation (Nadaf et al. 2011).

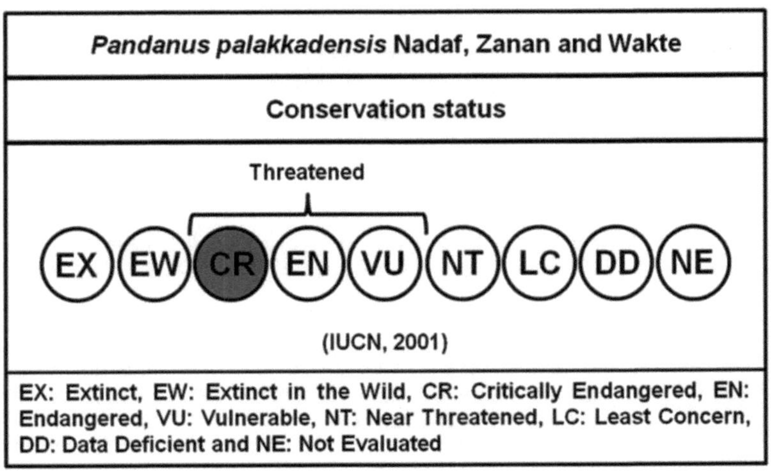

Pandanus mangalorensis

With an EOO of >100 km^2, an AOO of >10 km^2, and only one subpopulation, *P. mangalorensis* is assigned to Critically Endangered (CR B1ab [i, ii, iii]+B2ab [i, ii, iii]) according to IUCN Red List Categories and Criteria (IUCN 2001). The type locality is influenced by industrial activities and road constructions, and thus urgently demands conservation (Zanan and Nadaf 2012a).

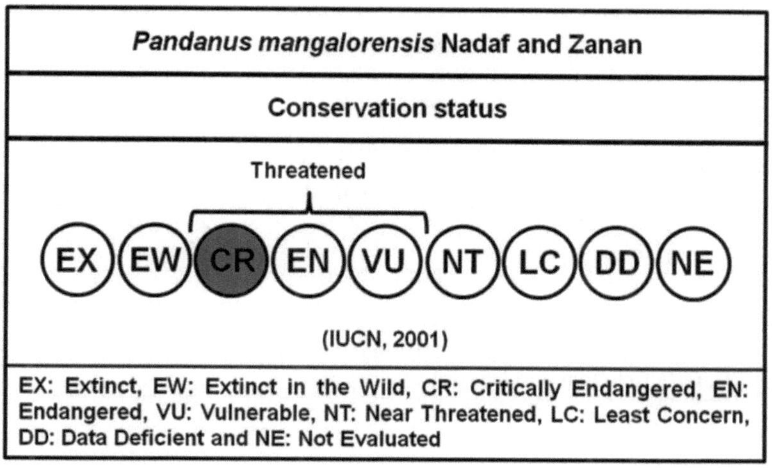

Pandanus nepalensis

Pandanus nepalensis is abundantly distributed in the Eastern Himalayan region in dense forest of cold highland areas. Based on distribution pattern and area, this species qualifies for the Least Concern (LC) category (IUCN 2001).

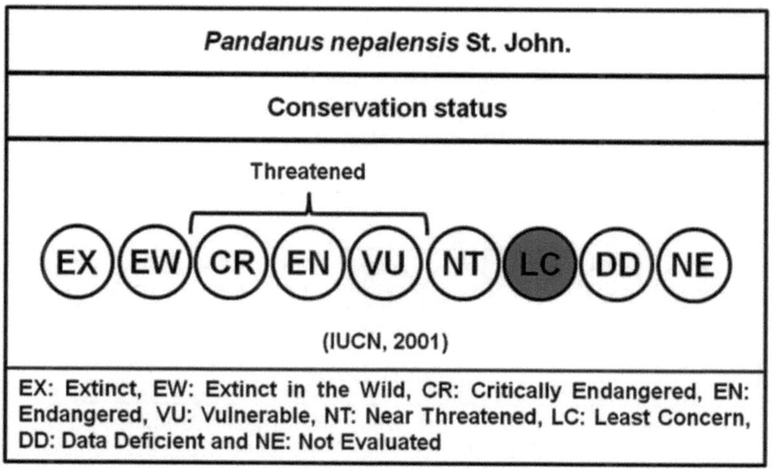

Pandanus emarginatus

Because the appropriate data are lacking, we have assigned this species under the Data Deficient category (DD).

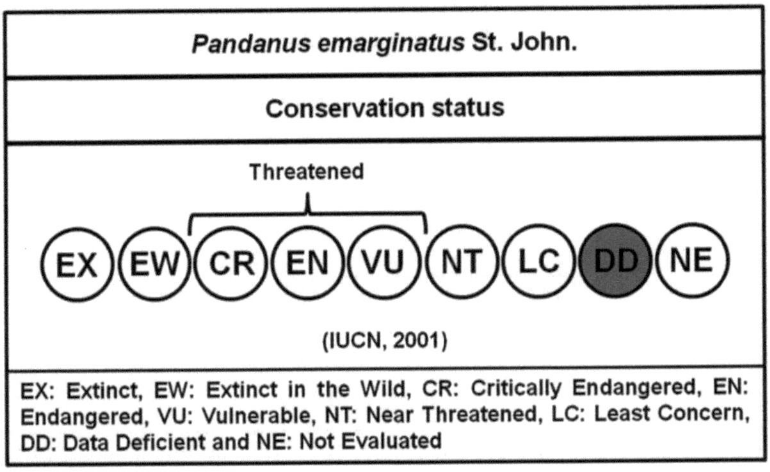

Pandanus diversus

Pandanus diversus is restricted to the Silchar district with an extent of occurrence (EOO) of 100 km^2 and an area of occupancy (AOO) of ca. 20 km^2. This species comprises six known subpopulations, two of which are located in deep forest areas of Koyla Godam village in the Silchar district and near the tea gardens of Dalu village, Silchar district. Based on IUCN Red List criteria (IUCN 2001), we have assigned the status of Vulnerable (VU D2), and therefore the species is considered to be facing high risk of extinction.

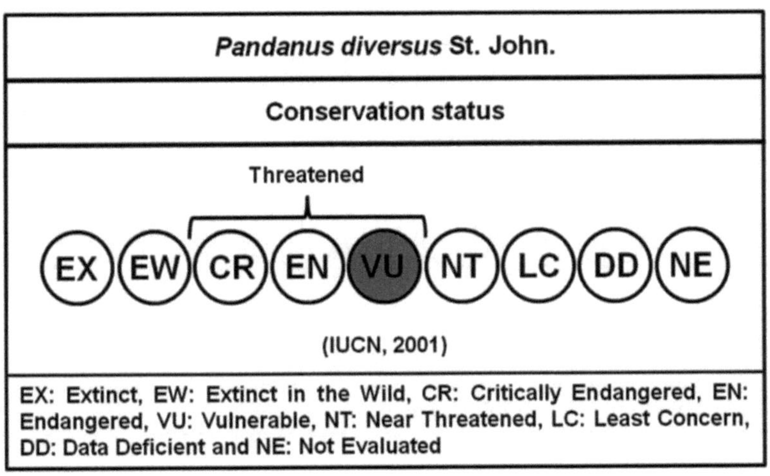

Pandanus leram

Pandanus leram is naturally distributed in the Andaman and Nicobar Islands (http://apps.kew.org). It was not evaluated using IUCN Red list categories and criteria and hence placed under the Not Evaluated (NE) category. A single individual is cultivated in the botanical garden of Botanical Survey of India, Yercaud, Salem district, Tamil Nadu.

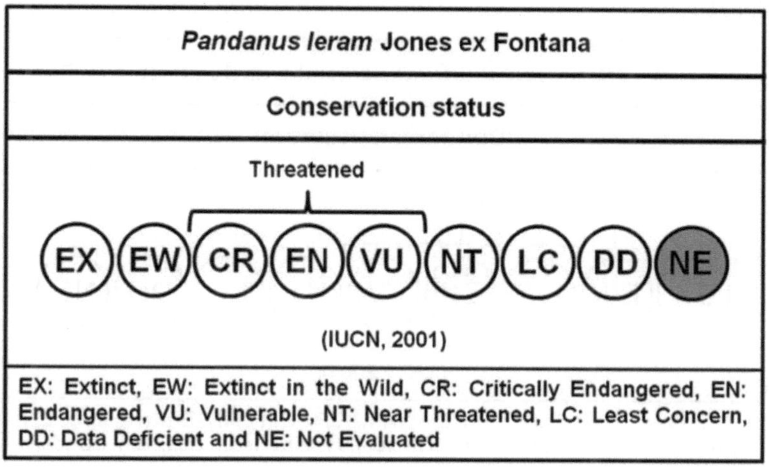

Pandanus unguifer

The species has an extent of occurrence (EOO) <100 km^2, an area of occupancy (AOO) <15 km^2, and comprises up to five known subpopulations, in which two locations are found in a conserved forest (Rambi Bazar, Darjeeling district) and others are near Cinchona Factory and in *Cinchona* plantation of Mungpoo village, Darjeeling district. As per the IUCN Red List criteria (IUCN 2001), the species is assigned to Endangered category (EN B1ab [i, ii, iii] + B2ab [i, ii, iii]).

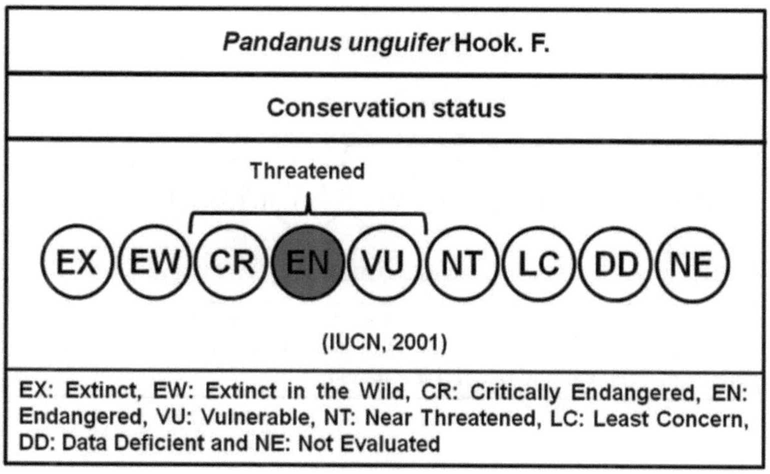

Pandanus martinianus

Pandanus martinianus has an extent of occurrence (EOO) of ca. 300 km^2, an area of occupancy (AOO) of 90 km^2 and comprises up to ten known subpopulations, three of which are found in a conserved forest (Arunachal Pradesh). *P. martinianus* should be considered as Endangered c [EN B1ab (i, ii, iii) + B2ab (i, ii, iii)] according to IUCN Red List Categories and Criteria (IUCN 2001; Zanan and Nadaf 2012b).

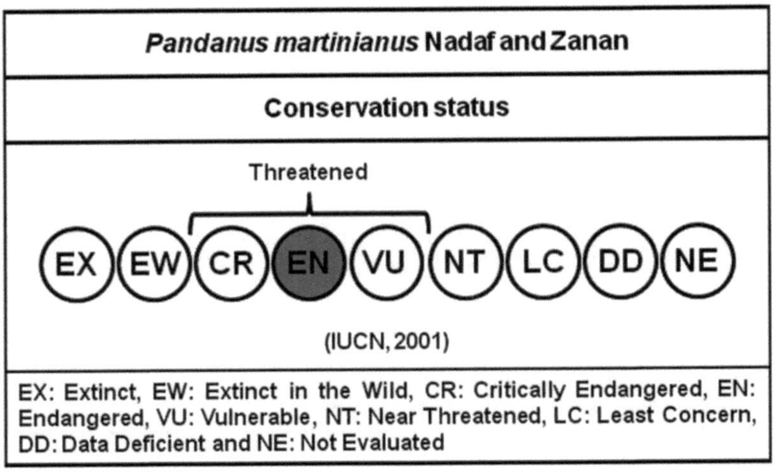

Pandanus kaida

Pandanus kaida is abundantly distributed in open forests of Western Ghats of India. Based on IUCN Red List criteria (IUCN 2001), we assigned this species under the Least Concern (LC) category.

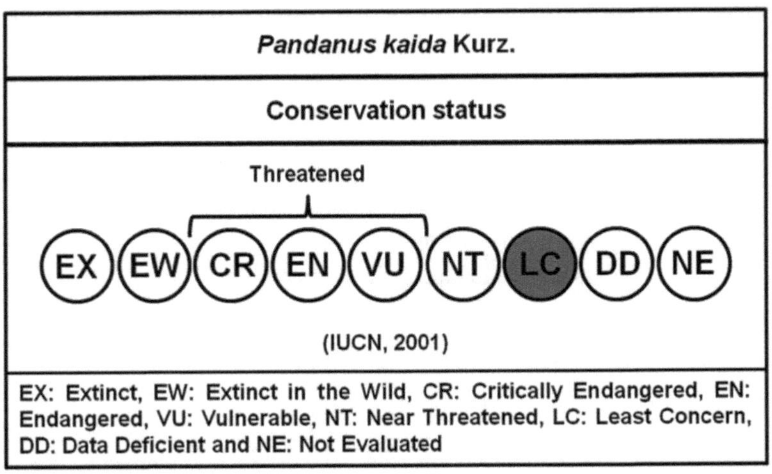

Pandanus odorifer

Pandanus odorifer is abundantly distributed in coastal region of west and east coast of India in open forest having warm humid conditions. Based on IUCN Red List criteria (IUCN 2001) we have assigned this species under Least Concern (LC) category.

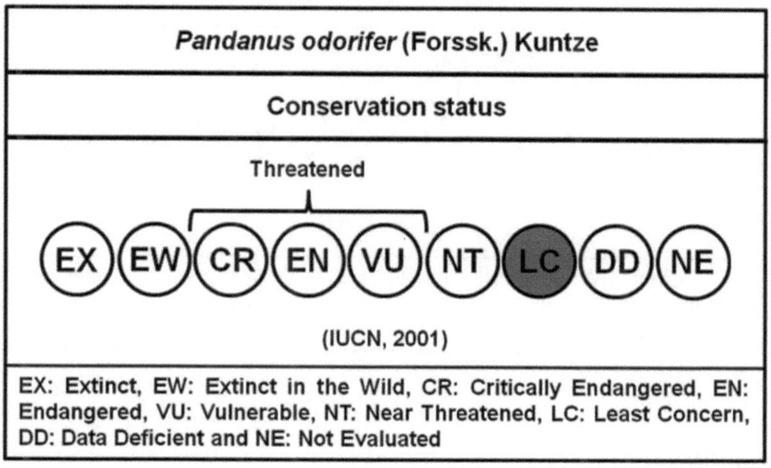

Pandanus dubius

Pandanus dubius is an exotic species to India and found only in gardens as a cultivated species, hence the conservation status is not determined.

Pandanus amaryllifolius

Pandanus amaryllifolius is known to survive only under cultivated conditions (Stone 1978; Wyk 2005; Wakte et al. 2009), hence the conservation status is not determined.

Thus, the study indicated that among the 16 Indian Pandanaceae members, 6 species are threatened, with categories of Critically Endangered (*P. palakkadensis* and *P. mangalorensis*), Endangered (*P. unguifer* and *P. martinianus*), and Vulnerable (*P. diversus* and *B. thwaitesii*). Six other species fall under the Least Concern category (*P. furcatus, P. unipapillatus, B. foetida, P. odorifer, P. kaida,* and *P. nepalensis*), *P. emarginatus* falls under the Data Deficient category and *P. leram* under the Not Evaluated (NE) category. The remaining two species, *P. amaryllifolius* and *P. dubius,* are exotic. We recommend six threatened species to include in the IUCN Red List (Table 8.1).

Benstonea thwaitesii

Benstonea thwaitesii is found in deep forest with humid climatic conditions in Western Ghats. It has an extent of occurrence (EOO) is <20,000 km^2, an area of occupancy (AOO) is <2,000 km^2, and comprises up to 25 known fragmented populations, the maximum of which are found in sacred groves of Maharashtra, Goa, and some are found in reserved forests of Karnataka and Kerala.

Based on IUCN Red List criteria (IUCN 2001), we assign as vulnerable category (VU) (VU B1ab [i, ii, iii] + B2ab [i, ii, iii]).

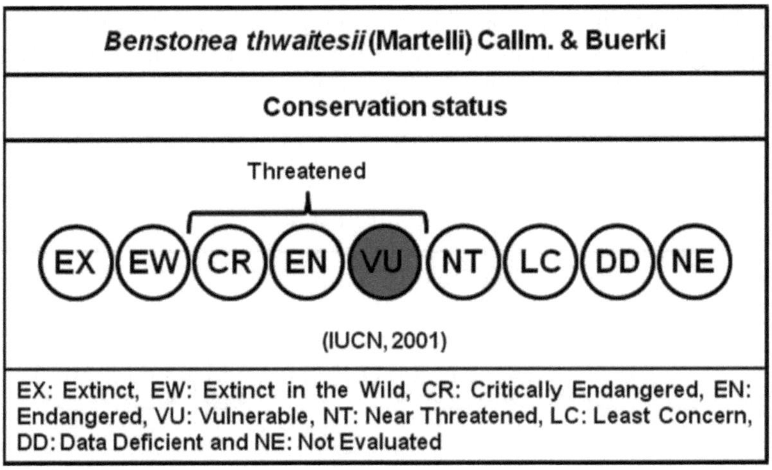

Benstonea foetida

Benstonea foetida is widely distributed in conserved forest regions. Based on IUCN Red List criteria (IUCN 2001), this species is assigned under the Least Concern (LC) category.

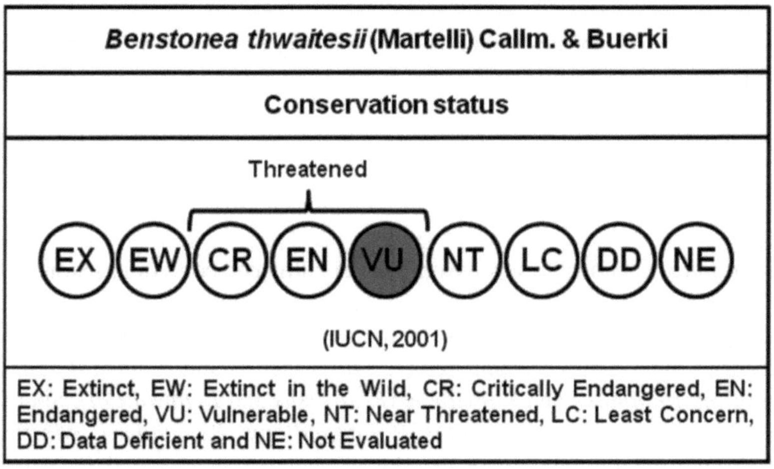

Table 8.1 Possible reasons of decline in population size of threatened species

Species	IUCN status	Reasons
P. palakkadensis	Critically endangered	Illicit cutting, shifting cultivation practices
P. mangalorensis	Critically endangered	Shifting cultivation practices, construction and widening of roads, industrial activities
P. unguifer	Endangered species	*Cinchona* plantation in the niche area of *P. unguifer*, deforestation
P. martinianus	Endangered species	Illicit cutting by tribals, shifting cultivation practices
P. diversus	Vulnerable	Deforestation, shifting cultivation practices, road construction and mining, extension of tea gardens
B. thwaitesii	Vulnerable	Illicit cutting for local use

Proposed Measures of Conservation for Threatened Species

The following measures are recommended to conserve the threatened species:

1. Population surveys and assessments and database creation
2. Improved protection efforts and a landscape approach to conservation
3. *In situ* approach through establishment of protected area network with appropriate management practices, corridors to link fragments, restoration of degraded habitats within and outside protected areas
4. *Ex situ* approach through collection of germplasm and maintaining in botanical gardens, in vitro regeneration, deposition of DNA in DNA bank
5. Reduction of biotic (anthropogenic) pressure by cultivating them in other localities
6. Regular population-habitat viability and risk assessment
7. Establishment of species preservation plots
8. Legal measures
9. Increased public awareness through educational programs, incentive programs, and volunteer monitoring programs.

References

Baillie JEM, Hilton-Taylor C, Stuart SN (eds) (2004) 2004 IUCN Red List of threatened species. A global species assessment. IUCN, Gland/Switzerland/Cambridge, pp xxiv+191

Batanouny KH (1991) Vegetation of the Summan (Arabia): pattern and process as affected by human impact and modern technology. In: IVth international rangeland congress, Montpellier. 4, pp 310–314

Biological diversity Act (2002) Ministry of law and justice. Legislative Department, New Delhi, 5 Feb 2003/Magha 16, 1924 (Saka)

Callmander MW, Schatz GE, Lowry PP II, Laivao MO, Raharimampionona J, Andriambololonera S, Raminosoa T, Consiglio T (2007) Application of IUCN Red List criteria and assessment of Priority Areas for Plant Conservation in Madagascar: rare and threatened Pandanaceae indicate new sites in need of protection. Oryx 41(2):168–176

Conservation International. http://www.conservation.org/where/priority_areas/hotspots/hotspots_revisited/Pages/hotspots_revisited.aspx

Heady HF (1963) Grazing resources and problems. A report to the government of Saudi Arabia. FAO, Rep. no.1614, Rome

IUCN (2001) IUCN Red List categories and criteria: version 3.1. IUCN species survival commission. IUCN, Gland, Switzerland & Cambridge, United Kingdom pp ii+33

Khoshoo TN (1996) India needs a national biodiversity conservation board. Curr Sci 71:506–513

Manilal KS (2001) Systematics and biodiversity research in India. Curr Sci 81:133–136

MoEF (2009) State of environment report India-2009. Environmental Information System (ENVIS), Ministry of Environment & Forests Government of India

Nadaf AB, Zanan RL, Wakte KV (2011) A new endemic species of Pandanaceae from India: *Pandanus palakkadensis*. Kew Bull 66:1–5

Nghia NH (2001) Conservation of forest genetic resources in Vietnam with reference to endangered tree species. In: Koskela J, Appanah S, Pedersen AP, Markopoulos MD (eds) Proceedings of the south east Asian moving workshop on conservation, management and utilization of forest genetic resources, 25 Feb, 10 Mar 2001, Thailand

Singh JS, Singh SP, Gupta SR (2008) Ecology, environment and resource conservation. Anamaya Publishers, New Delhi, pp 517–553

Stone BC (1978) Studies in Malesian Pandanaceae XVII. On the taxonomy of 'Pandan Wangi': a Pandanus cultivar with scented leaves. Econ Bot 32(3):285–293

Wakte KV, Nadaf AB, Thengane RJ, Jawali N (2009) *Pandanus amaryllifolius* Roxb. cultivated as a spice in coastal regions of India. Genet Resour Crop Evol 56:735–740. doi:10.1007/s10722-009-9431-5

Wyk BEV (2005) Food plants of the world: identification, culinary uses and nutritional value. Times Editions–Marshall Cavendish, Singapore, p 275

Zanan RL, Nadaf AB (2012a) *Pandanus mangalorensis*: a new species of Pandanaceae from southern India. Kew Bull 67(3):555–559

Zanan RL, Nadaf AB (2012b) *Pandanus martinianus* (Pandanaceae), a new species endemic from North Eastern region of India. Phytotaxa 73:1–7

About This Book

Indian screw pine family Pandanaceae represents three genera, in which genus *Pandanus* and *Benstonea* are distributed in two hotspots in India – the Western Ghats and the Northeast Himalayan region. For the first time, Indian Pandanaceae has been assessed for its taxonomic status and phylogenetic relationship. The extensive survey by the authors led to the discovery of three new *Pandanus* species (two from the Western Ghats and one from the Northeast Himalaya). The present taxonomic revision confirmed total number of *Pandanus* species to 14 that are distributed in the Southern India (9 species) and Northeast Himalayan region (5 species). Genus *Benstonea* is represented by two species, one from Southern India and another species common to both regions. A detailed species identification key is given along with conservation status of each species following IUCN Red List Categories and Criteria (2001). The study revealed that, out of 16 *Pandanus* and *Benstonea* species, six species are under threatened categories.

In recent years, a chloroplast DNA-based molecular phylogenetic approach has been followed to understand the evolutionary relationship among the plant species. The interrelationship among the 14 *Pandanus* species at infrageneric level has been worked out using this approach, which has led to the rearrangement of some species to the subgenera proposed by Stone (1974). Moreover, the close relationship between *Pandanus* and *Benstonea* has been confirmed and the interrelationship of Indian *Pandanus* genus in global context is given. This book also describes the economic importance of each *Pandanus* species.

A. Nadaf and R. Zanan, *Indian Pandanaceae - an overview*,
DOI 10.1007/978-81-322-0753-5, © Springer India 2012

About the Authors

Dr. Altafhusain Nadaf is a Senior Assistant Professor at the Department of Botany, University of Pune, Pune, India, and has worked in the area of biochemistry and molecular genetics of Indian Pandanaceae for the past 10 years. He was awarded the DST-BOYSCAST Fellowship to work as a visiting fellow at the Centre for Plant Conservation Genetics, Southern Cross University, Lismore, Australia, for one year in 2010–2011. He has received research grants from several funding agencies, including the DST-Fast Track Scheme for Young Scientists and the DBT-Rapid Grant for Young Investigators (RGYI). He has successfully guided several Ph.D. and M.Phil. students and women scientists (under DST-Women Scientists Scheme – A & B). He has also worked as a reviewer for many international journals. He has to his credit more than 40 research papers published in peer-reviewed national and international journals and has presented his research work at the national and international level.

Dr. Rahul Zanan is a young researcher and has worked on the taxonomy and phylogeny of Indian Pandanaceae for the past 5 years. He obtained his doctoral degree from University of Pune, Pune, India. He has published eight research papers in international journals and is the authority on three new *Pandanus* species of India. He has received several awards for best oral and poster presentations at international and national conferences.

A. Nadaf and R. Zanan, *Indian Pandanaceae - an overview*,
DOI 10.1007/978-81-322-0753-5, © Springer India 2012

Glossary

Abaxial On the side away from the axis or turned away from the axis

Abonic Hard

Abrupt Suddenly narrowed at the apex

Adaxial On the side towards the axis or turned towards the axis

Aerial root A root that develops from a location on a plant above the surface of the earth or water, as from a stem

Aggregate fruit Formed from free pistils of a single flower which are coherent in fruit

Ancestors The actual or hypothetical organism or stock from which later kinds evolved

Androphore A stalk or column supporting the stamens formed through fusion of filaments

Angiosperm Phylogeny Group (APG) An informal international group of systematic botanists who came together to establish a consensus on the taxonomy of flowering plants through phylogenetic studies

Arboreous Resembling a tree in form and branching structure, distinct from a shrub

Area of occupancy (AOO) The area within its extent of occurrence, which is occupied by a taxon, excluding cases of vagrancy

Attenuate Tapering to a point or gradually narrowing

Basifixed Attached by the base

Berry The most generalized type of fleshy fruit derived from a single ovary and with the pericarp wholly fleshy

Bifid Divided into two lobes

Bifurcate Forked; divided into two branches

Biodiversity The variability among living organisms on the earth, including the variability within and between species and within and between ecosystems

Biogeographic regions The regions of the world containing recognizably distinct and characteristic endemic fauna or flora

Biogeography The study of the geographic distribution of plants and animals

Bootstrapping analysis A way to judge the strength of support for nodes on phylogenetic trees

Bracts Any more-or-less reduced or modified leaf associated with a flower or an inflorescence, but not part of flower itself

Canopy The branches and leaves of woody plants that are formed some distance above the ground

Carpel The fertile leaf of an angiosperm, which bears the ovules

Cephalia Fruit

Characters The abstraction of an observable physical or biochemical trait of an organism

Chemosynthesis The formation of organic material by some bacteria by means of energy derived from chemical change, in contrast to transfer of energy by complex chemical reaction as in respiration

Chloroplast DNA The DNA present in the chloroplast

A. Nadaf and R. Zanan, *Indian Pandanaceae - an overview*,
DOI 10.1007/978-81-322-0753-5, © Springer India 2012

Clade Any individual evolutionary lineage from beginning to end

Clavate Club-shaped

Cluster Grouping of different organisms

Coastal zone A spatial zone where interaction of the sea and land processes occurs

Compound Made up of several similar parts

Concave Hollowed out; like a saucer

Concentric Having a common center

Consensus tree The agreement between two or more phylogenetic trees

Conservation Conservation is the science of the protection and management of biodiversity.

Coriaceous Leathery texture

Cretaceous The period of geological time running from 135 to 63 million years ago

Critically Endangered (CR) A taxon is Critically Endangered when the best available evidence indicates that it meets any of the IUCN criteria A to E for Critically Endangered, and it is therefore considered to be facing an extremely high risk of extinction in the wild.

Cuneinate Attached to each other or bring to close

Cylindrical Having the shape of a cylinder

Data Deficient (DD) A taxon is Data Deficient when there is inadequate information to make a direct, or indirect, assessment of its risk of extinction based on its distribution and/or population status.

Deccan Peninsular zone The geographical region is bounded by the Arabian Sea in the west, the Indian Ocean in the south, and the Bay of Bengal in the east.

Decumbent Lying flat or being prostrate but having the tip growing upwards

Deltoid Point at the apex

Denaturation Denaturation is the process by which double-stranded deoxyribonucleic acid (DNA) unwinds and separates into single-stranded strands through the breaking of hydrogen bonding between the bases.

Dioecious Producing male and female flowers on separate individuals

Divaricately Branching or spreading widely from a point or axis

Drupe A fleshy fruit with a firm endocarp that permanently encloses the usually solitary seed or with separate portions of the endocarp enclosing each of two or more seeds

Edible Fit to be eaten as food; eatable

Ellipsoid A three-dimensional geometric shape resembling a flattened sphere

Endangered (EN) A taxon is Endangered when the best available evidence indicates that it meets any of the IUCN criteria A to E for Endangered, and it is therefore considered to be facing a very high risk of extinction in the wild.

Endemic Native to one geographic locality and nowhere else

Endocarp The inner layer of the fruit wall, often hard, fibrous, stony, or otherwise resilient.

Ensiform Shaped like a sword blade

Eocene The epoch of geologic time from 40 to 58 million years ago

Epidermis The outer single layer of cells of the primary plant body, usually consisting of a single layer but sometimes several layers thick

Erect Set at right angles to the organ from which it arises

Evergreen Retaining leaves throughout the year

Evolution A gradual change by which genetic changes have taken place in populations of animals and plants over millions of years in response to environmental changes

Exotic Introduced from another place; not native to the locality where presently found

Extent of occurrence (EOO) Extent of occurrence measures the spatial spread of the areas currently occupied by the taxon.

Extinct (EX) A taxon is Extinct when there is no reasonable doubt that the last individual has died.

Extinct in the Wild (EW) A taxon is Extinct in the Wild when it is known only to survive in cultivation, in captivity, or as a naturalized population well outside the past range.

Family All genera that share common characteristics form a "family."

Fibrous With loose, woody fibers, and thereby often tough

Filament The stalk of the stamen, terminated by the anther

Fleshy Thick and soft but not juicy

Flora A descriptive list of plants of an area, including key of identification

Forked Divided or separated into two branches

Fossils The remains of organisms from past geological ages preserved in sedimentary rocks either as actual structures or as impressions, casts, or molds

Fragrance Pleasant odor

Fusiform Elongated and tapering towards each end

Gangetic plains Fertile lowland region of north-central India

Gene Pool The total number of genes of every individual in an interbreeding population

GenBank A database of nucleic acid and protein sequences at the National Library of Medicine in the United States of America, compiled from international sources

Genetic diversity Genetic variation between and within species, which is measured by determining the proportion of polymorphic loci across the genome

Genetic resources Genetic resources are genetic material of plants, animals, or microorganisms of value as a resource for future generations of humanity.

Geographical distribution The natural arrangements of animals and plants in particular regions

Glabrous Without hairs or glands

Glaucous A surface coated with a whitish bloom that rubs off when handled

Globose Spherical, a perfect three-dimensional circle

Gondwanaland The hypothetical protocontinent of the Southern Hemisphere that, according to the theory of plate tectonics, broke up into Asia, Australia, Antarctica, Africa, and South America

Gradually tapering Gradual decrease

Guard cells A pair of reniform cells bounding a stoma

Homologues Structures that are similar or may be structurally or functionally different but that have descended from a common type

Illustrations A visual representation in terms of a picture or diagram that is used make some subject more easier to understand

Inflorescences The flower-bearing part of a plant, usually with two or more flowers but sometimes reduced to a single flower

Infrageneric Produced between two or more species belonging to the same genus

Infructescence The grouping or arrangement of fruits on a plant

Intergenic spacer The region between two genes

Intron A segment of the primary transcript of a eukaryotic gene

Islands Any piece of sub-continental land that is surrounded by water

IUCN Red List Category Code for the determination of Threatened Status of a species using the 2001 IUCN Red List Categories and Criteria

IUCN Red List Criteria Codes referring to the specific IUCN Red List Criteria that apply to the species and justify its inclusion in its assigned IUCN Red List Category

Key A tool used in plant or animal identification

Lanceolate Shaped like the point of a lance, much longer than broad, widening above the base and tapering to the apex

Landward Toward the land

Lateral subsidiary cell Cells adjacent to lateral side of guard cells those are distinct in appearance from ordinary epidermal cells

Leaf scar Scar left on a twig when a leaf is lost

Leafy bract Leaf-like bract

Least Concern (LC) A taxon is of Least Concern when it has been evaluated against the IUCN Criteria and does not qualify for Critically Endangered, Endangered, Vulnerable, or Near Threatened

Leathery Texture of leather, thick yet somewhat flexible, tough

Liana A woody, usually high-climbing vine, often from tropical regions

Linear Long and narrow, with parallel sides

Linear-lanceolate Intermediate between linear (elongated, and with parallel side) and lanceolate (narrow, as a lance, much longer than wide, with the widest point below the middle)

Linear-oblong Intermediate between linear (elongated, and with parallel side) and oblong (two to four times longer than broad with nearly parallel sides)

Locule A seed cavity in an ovary or fruit; a compartment in any container

Macrofossil A fossil that is large enough to be studied directly, without the aid of a microscope

Malabar plains A region of Southwest India bordering the Arabian Sea and bounded on the east by the Western Ghats

Margin Edge of leaf

Median Divides an organism or organ into symmetrical parts

Mesocarp The middle layer of a fruit wall, which may be dry and fibrous, fleshy and juicy, or oily

Mesozoic The geological era, this spans from the Triassic, Jurassic, and Cretaceous periods, from 245 to 65 million years ago

Micromorphology The fine-level structures or morphology of an organism component visible through microscopy

Midrib The largest vein running down the middle of a leaf base to the leaf tip, often dividing the leaf into similar halves

Miocene A geological era lasting from approximately 25–5 million years ago

Mitochondrial genome DNA present in mitochondria

Molecular phylogeny Use of molecular data (DNA, RNA, and/or proteins) as characters for phylogenetic analysis

Monogeneric The sole member of the group

Monophyletic A group of organisms that are assumed to have originated from the same ancestor

Mya Million years ago

Near Threatened (NT) A taxon is Near Threatened when it has been evaluated against the IUCN criteria but does not qualify for Critically Endangered, Endangered, or Vulnerable now, but is close to qualifying for or is likely to qualify for a threatened category in the near future.

Neotype A specimen selected as the type subsequent to the original description in cases in which the primary types are definitely known to have been destroyed

Not Evaluated (NE) A taxon is Not Evaluated when it is has not yet been evaluated against the IUCN criteria.

Oblanceolate Leaf having a broad rounded apex and a tapering base

Oblong Lozenge-shaped, longer than broad with parallel sides and rounded ends

Outgroup taxa In cladistics, a taxon that is not part of the clade under consideration, but is including in the analysis in order to provide a baseline

Ovate Egg-shaped in two-dimensional outline, the attachment at the broad end

Ovoid Egg-shaped in three-dimensional form

Palaeontology The study of fossil

Paleocene Earliest epoch of the Tertiary period, spanning the time between 65 and 55.5 million years ago

Paleotropical A biogeographic region that includes the tropical or subtropical regions of the Old World

Palynology The study of identifying plant spores and pollen

Papillae Extension of epidermal cell as a small outgrowth

Paraphyletic group The groups that have a common ancestry but that do not include all descendants

Peduncle The stalk of a cluster of flowers, or of a single flower

Pendulous Hanging downward

Peninsular India The area encompassing states of Andhra Pradesh, Karnataka, Kerala, and Tamil Nadu as well as the union territories of Lakshadweep and Pondicherry

Perianth Collective term for sepals and petals (calyx + corolla), especially when these are not easily distinguishable from one another

Pericarp The mature ovary wall of a fruit

Phalanges A group of fused carpels that separates as a unit from other such units, all of which comprise a compound fruit

Phylogenetic tree A branching tree-like, diagrammatic representation of the evolutionary relationships and patterns of branching in the history of the organisms being considered

Phylogeny The evolutionary history of an organism

Pileus The cap of an ogaric, bearing the hymerinium on its lower surface

Pyramidal Resembling a pyramid

Pistillode A sterile pistil, sometimes present in staminate flowers

Polymerase chain reaction (PCR) A procedure that allow the production of multiple copies of a specific DNA sequence, provided that the base pair sequence of each end of the target is known

Polytomy In a cladistic phylogeny, a node where more than two lineages descend from a single ancestral lineage

Polytypic Species with considerable phenotypic variation

Predominant species Most frequent or common growing species

Prickles A sharp-pointed outgrowth from epidermal layers of any plant part

Primers A short oligonucleotide sequence

Prop roots Adventitious roots growing from the trunk or branches of a woody plant or the stem of an herb meant to anchor and support the plant

Prostrate Horizontally flat or procumbent stem

Protein The macromolecule composed of one or more polypeptides, each comprising a chain of amino acids linked by peptide bonds

Provinces An area with respect to its flora, fauna, or physical characteristics

Racemose Pertaining to a broad class of inflorescences characterized by flowering in acropetal sequence

Retuse With a small terminal notch in an otherwise rounded or obtuse apex

Rigid Physically inflexible or fixed

Sect. Section

Semiorbacular Slightly circular or approximately round at the apex

Sequences The linear order of nucleotide along a DNA or RNA molecule and the process of obtaining this

Sheath A more or less tubular structure surrounding a plant part

Sister group Taxa or clades that share common ancestor

Solitary Flowers occurring one in each axil

Spadix The inflorescence in aroids, a thick, often fleshy spike that is surrounded or subtended by a bract

Spathe The bract(s) surrounding or subtending a flower cluster or spadix

Spherical Globose, round like a basketball

Spike An unbranched inflorescence like a raceme having sessile flowers

Spine A sharp-pointed part derived from a leaf petiole, midvein, and lateral vein or stipule

Stamen The pollen-bearing organs of seed plants

Staminate inflorescence A unisexual flower in which stamens are present and fertile while the pistil is vestigial or absent

Stamenophore A stalk supporting to the stamens

Subg. Subgenus

Subgenera A taxonomic category that ranks below genus and above species

Subspecies A taxonomic category lower than a species, usually distinguished by its morphology, physiology, and often by its geographic isolation

Subtropical The region bordering the tropics, usually between ~20° and 40° latitude in both hemispheres, adjacent to the tropics with warm temperatures and little rainfall

Subulate Wedge-shaped, broadest at the base, tapering gradually toward the apex

Syncarpus Within the carpels united to form a compound pistil

Synonym A name rejected in favor of another because of nomenclatural technicalities or an earlier date of publication

Terminal subsidiary cells Cells adjacent to terminal side of guard cells those are distinct in appearance from ordinary epidermal cells

Terrestrial Growing on land

Triangular Having three sides, the attachment on one of the sides

Trigonous Having three sides or faces

Tropical Regions of the earth lying between the Tropic of Cancer and the Tropic of Capricorn extending around the equator where the temperature and humidity are high

Tubuliform Synonym for tubular, shaped like a tube

Umbellate Inflorescence type in which all the flower pedicels arise from the same point at the apex of the main axis

Unarmed Lacking prickles, spines or thorns

Western Ghats A low mountain range in Western Peninsular India, parallel to the Malabar coast of the Arabian Sea extending from Gujarat to Kerala; about 1,600 km long

Woody Tissues that are hard due to the presence of lignin

Index

A. Nadaf and R. Zanan, *Indian Pandanaceae - an overview*,
DOI 10.1007/978-81-322-0753-5, © Springer India 2012

MIX
Papier aus verantwortungsvollen Quellen
Paper from responsible sources
FSC® C105338

If you have any concerns about our products,
you can contact us on
ProductSafety@springernature.com

In case Publisher is established outside the EU,
the EU authorized representative is:
Springer Nature Customer Service Center GmbH
Europaplatz 3, 69115 Heidelberg, Germany

Printed by Libri Plureos GmbH
in Hamburg, Germany